Living on the Grid

Living on the Grid
The Fundamentals of the North American
Electric Grids in Simple Language

WILLIAM L. THOMPSON

LIVING ON THE GRID
The Fundamentals of the North American Electric Grids in Simple Language

Copyright © 2016 William L. Thompson.

All rights reserved. No part of this book may be used or reproduced by any means, graphic, electronic, or mechanical, including photocopying, recording, taping or by any information storage retrieval system without the written permission of the author except in the case of brief quotations embodied in critical articles and reviews.

iUniverse books may be ordered through booksellers or by contacting:

iUniverse
1663 Liberty Drive
Bloomington, IN 47403
www.iuniverse.com
1-800-Authors (1-800-288-4677)

Because of the dynamic nature of the Internet, any web addresses or links contained in this book may have changed since publication and may no longer be valid. The views expressed in this work are solely those of the author and do not necessarily reflect the views of the publisher, and the publisher hereby disclaims any responsibility for them.

Any people depicted in stock imagery provided by Thinkstock are models, and such images are being used for illustrative purposes only.
Certain stock imagery © Thinkstock.

ISBN: 978-1-4917-9043-4 (sc)
ISBN: 978-1-4917-9045-8 (hc)
ISBN: 978-1-4917-9044-1 (e)

Library of Congress Control Number: 2016903046

Print information available on the last page.

iUniverse rev. date: 05/20/2016

An analysis of the North American electric grid: what it is, who owns it, how it is controlled, what keeps it going, and what happens when it blacks out. Written by an electrical engineer who was in the business for more than thirty-eight years and who has tried to simplify this complex topic for the person who is curious about it and also for those who are new to the business.

To all the operators working seemingly never-ending shifts to keep the lights on for the rest of us. You are an amazing group of people.

CONTENTS

List of Illustrations xi

Foreword xiii

Preface xv

Acknowledgments xvii

Chapter 1: The Grid 1
   Introduction to the Grid 1
   Starting with a Small Electric System 3
   The Difference between Power and Energy 4
   The Growth of Small Electric Systems 5
   The Next Step in Grid Development: Interconnecting 7
   Connecting the Grid in a Network 9
   Alternating Current vs. Direct Current 11
   Building an Interconnected System 13
   Texas: State vs. Federal Jurisdiction 15
   An Analogy for an Interconnected Grid 16
   Grid Evolution Was Not Random 17
   Ownership of the Grid 18
   Another Technical Concept: Voltage 20
   The Three Electrical Phases 21
   How Electricity Flows through the Grid to a Home 23
   Bad Things Can Happen to the Grid 25
   Stories from a Control Room 26

Chapter 2: Grid Control 29
Introduction to Control of the Grid 29
Why Does the Grid Need to Be Controlled? 30
- Reactive Power 31
- Generation and Load Balance 32
- Storms or Equipment-Failure Events 33
- Generation Loss 35
- System Design / Cascading Events 35
- Stability 36
- Markets 37
- Solar Magnetic Disturbances 39

What Can We Do to Keep the Grid Under Control? 40
- Establish Design Rules and Build the Grid 40
- Establish Operating Rules and Operate the Grid 44
  - Rules for Balancing Generation and Load 44
  - Operating to the Single-Contingency Rule 52
  - Establish Regional Entities That See the Big Picture 55
  - Other Operational Rules 58
- Establish Rules for the Market and for Good Performance by Wholesale Purchasing/Selling Entities 59
- Establish Rules for Generators and for Good Performance by Generators 63
- Utilize Quick-Acting Automatic Protection Systems 64
- Ensure Maintenance 66

Chapter 3: Reliability Tools Used by Operators in Real Time 68
Introduction to the System Operator's Tools 68
Operational Tools to Maintain the Grid 69
- Redispatching Generation 69
- Reconfiguring the Grid 70
- Using Load-Management Programs 72
- Reducing Voltage 73
- Making Public Appeals 74
- Instituting Rotating Blackouts 75
- Doing Nothing 78

Information Available for the Operator 79

Chapter 4: The Government's Role in the Grid 83
   Introduction 83
   Federal Energy Regulatory Commission (FERC) 84
      The Energy Policy Act of 1992 (EPAct) 85
      FERC Orders 888 and 889 86
         Open Access to Transmission 87
         Generation Transactions 88
         Separation of Affiliated Businesses 89
         Unintended Consequences 89
      The Energy Policy Act of 2005 (EPAct05) 92
      FERC Order 693 and Additional Orders on Standards 93
   One Critical Federal Issue: The Independent System Operator 94
   Department of Energy 98
   State Regulation 98
   North American Electric Reliability Corporation (NERC) 102
   Other Entities That Impact the Grid 104

Chapter 5: Generation 106
   Introduction 106
   General Terms and Concepts 107
      Economic Terms and Concepts 107
      Technical Terms and Concepts 109
   Nuclear Generation 113
   Coal Generation 118
   Natural Gas Generation 121
   Oil Generation 122
   Hydro Generation 123
   Wind Generation 124
   Solar Generation 128
   Geothermal Generation 129
   Fuel Cells 130
   Energy Storage 131
   What Is Green Energy? 132
   What Is the Best Mix of Generation? 133
   Cost-Based vs. Price-Based System: The Market 139

Chapter 6: Competition: The Market 140
    Introduction to the Wholesale Market for Electricity 140
    Purchases and Sales of Wholesale Electricity the Old Way 141
    The Difference between Cost and Price 145
    The Products Traded at the Wholesale Level 146
    Changes Initiated by the Open Access Transmission Tariff 148
    Markets Managed by Independent Organizations 152
    Generation or Transmission Built by Independents 152

Chapter 7: Grid Blackouts 154
    Introduction to the Blackout Discussion 155
    The History of Blackouts 155
    The Northeast Blackout of 2003 156
    Setting the Stage for a Blackout 157
    It Takes a Combination of Big Mistakes to Cause a Blackout 158
    Tools, Trees, Training: Lessons Learned 160

Chapter 8: The Future of the Electricity Business 162
    Introduction 162
    The Death Spiral and Why It Won't Happen 163
    Net Metering: Another Form of Subsidizing Renewables 165
    Microgrids and Other Ideas 166
    The Smart Grid 168
    Cybersecurity 171
    Terrorism and the Grid 172
    A Future View 174

Glossary of Terms 179

Suggested Reading and References 187

Index 191

LIST OF ILLUSTRATIONS

Chapter 1
Diagram 1-1: The Three Grids of North America....................2
Diagram 1-2: A Grid in Your Home......................................3
Diagram 1-3: An Early Electric System 6
Diagram 1-4: A Slightly Improved System...........................7
Diagram 1-5: Early Electric System Tie-Lines......................8
Diagram 1-6: Radial Electric System 10
Diagram 1-7: Network Electric System 10
Diagram 1-8: Alternating Current.......................................12
Diagram 1-9: Direct Current Tie-Lines between Grids.....14
Diagram 1-10: Water Barrel Analogy for the Grid 16
Diagram 1-11: Transformers Allow the Grid to Connect Together 21
Diagram 1-12: The Three Phases of Alternating Current.................22
Diagram 1-13: Three-Phase Transmission Line.....................23

Chapter 2
Diagram 2-1: Balancing Generation and Load 46
Diagram 2-2: Contract Path vs. Actual Flows 56

Chapter 5
Diagram 5-1: Load for a Typical Summer Peak Day 136
Diagram 5-2: Load for a Typical Winter Peak Day 137
Diagram 5-3: Load for a Typical Shoulder Season Day..................137
Diagram 5-4: Generation Mix for a Typical Summer Peak............139

FOREWORD

Why a book about the electric grid? Our society is dependent on our electrical infrastructure, yet most of us know little about the grid. The available books on the subject present highly technical descriptions of the electrical engineering aspects of the grid that the average person would not understand or, for that matter, read. This book has been written by someone who understands the technical aspects of the electric grid and who has endeavored to describe the main aspects for the general public. I believe that this book will fill a void by providing a simple description of the grid, including how the grid is controlled, how it is regulated, energy-generation alternatives that support it, and what the future of the grid may be.

There are many topics that are addressed in simple, plain language, far more topics than were relevant to the field of electrical engineering when I first encountered it. System economics, alternative fuels for electric generation, regulatory rules, environmental concerns, reliability, and wholesale competition are some of the topics addressed herein— and at a level that should be interesting and informative to many, including those of us not in the electricity business. The electric grid is a very interesting thing that is far more complicated than most people realize. After reading this book, one will have a much better appreciation for this infrastructure on which we all depend. The personal stories and simple analogies that the author presents will go a long way toward increasing a reader's understanding and, hopefully, enjoyment of the subject.

Cleve Moler, cofounder of Mathworks and first author of MATLAB

PREFACE

Our way of life in the United States is dependent on electricity, and the flow of electricity in the United States is dependent on the electric grid. However, most people have very little understanding of what the grid is or of what it does. I spent twenty years working inside a control room that was responsible for the operation of a significant part of the grid in Virginia and North Carolina. Over that period I showed our control center to hundreds, if not thousands, of people. Many people asked me to suggest a book they could read that would explain the grid in simple terms. The truth is that I have yet to see such a book for the general public, as most books that include discussions of the grid are far too technical for anyone but electrical engineers. My intent in writing this book is to meet the need for a simple explanation of the grid that both engineers and non-engineers can understand, leading them to appreciate what this huge interconnected machine does for them.

This book will answer basic questions and satisfy curiosity about the electric grid. I have used simple examples and diagrams, avoiding detailed explanations of the laws of physics as much as possible. To help with an understanding of the concepts presented, I have interjected stories of actual events. Wherever possible, I simplified my explanations to the extent that I left out complications that are not needed for an overall understanding. I have not attempted to address every facet of electrical engineering or to address every complexity of grid operating paradigms.

If someone is looking for a more technical description of the grid, or of the laws of physics related to electricity, then one may find many other books that provide such information.

I have presented generic data in a few places (such as the percentage of energy derived from coal) only to provide scale. Information such as this is readily available, so rather than use profuse footnotes, I have assumed that anyone can probe deeper if desired. There are many references in the "Suggested Reading" section to get someone started.

I am a retired electrical engineer with thirty-eight years of experience in the electric utility business and with several years of consulting experience after retirement. For the last twenty years of my career, I was the person in charge of grid operations at Dominion Virginia Power. The electric system at Dominion is one of the top ten in total size in the United States, with a peak load recently exceeding 21,000 megawatts. Over those twenty years, I experienced the need to order rotating blackouts throughout Dominion's Virginia–North Carolina territory, witnessed the Enron rise and fall, went through the Y2K scare, experienced the northeast blackout of 2003, witnessed the federal government's drive to create a competitive wholesale market, and saw attempts by the states to institute retail competition for electricity.

I should add that all the opinions, observations, and predictions in this book are my own and are not endorsed by Dominion Virginia Power. Dominion has not endorsed this book and is not providing any commission for the writing. And, of course, any errors are my own.

ACKNOWLEDGMENTS

First, I must thank all the people who helped me survive in the complex business of electricity delivery. They patiently explained to me how things work, and I truly appreciate that.

I offer special thanks to people who have helped with the development of this book, starting with Cleve Moler, who encouraged me to continue with the project and provided some very good advice, which I have tried to heed. I can't thank enough Monty Jackson, who spent a great deal of his own time reading and commenting on some of the early chapters and who encouraged completion of this effort. Then there is my sister, Ruth Walton, who read through some parts of the book and pointed out things that didn't make sense so I could rework them. My colleagues Mike Regulinski and George Marget advised me on the position of Dominion Virginia Power in light of my developing this book. John Lambert and Brian Garbera encouraged me to continue writing and gave me some great ideas about things to delve into. And last but certainly not least is my wife, Betty Thompson, who encouraged me, put up with my sitting at the computer when I should have been doing something else, proofread my drafts, and sketched most of the diagrams provided in this book.

Chapter 1

THE GRID

It seemed like a typical summer afternoon in the system operations control room in Richmond, Virginia, on August 14, 2003, but it wasn't going to be typical at all. Suddenly, the system operator on duty broke the calm by shouting something like "Holy smoke! There is no way the grid can stay together; it's going to crash!" Although our jobs as operators of the grid were to protect the grid in our area of Virginia and North Carolina, there was nothing we could do to prevent what was getting ready to happen. Within seconds, the largest grid blackout in US history occurred in the northeastern part of the United States and in a large part of Canada, leaving more than fifty million people without power. In Virginia and North Carolina, the power stayed on. We were fortunate, knowing how devastating it would be to the public to lose power across such a large area.

Introduction to the Grid

What happened? Why did such a large portion of the grid black out while other parts of the grid stayed in service? How did our operator know that the grid was close to collapsing? Who are the people that control the grid, and how is it controlled? And more fundamentally, what is the grid? These questions are extremely important to our society and our way of life since we are so dependent on electricity.

The grid is a huge connection of electric power lines, generators, transformers, and other electrical equipment. The grid exists to take the electric-power output from generators and then deliver that power to loads. The grid that we know today serves us extremely well, to the point that we have structured a large part of our lives around its existence. We rarely consider the consequences of total grid failure. And yet most people probably know more about what it means to "live off the grid" than what the grid is.

Actually, there are a number of grids in the world today. In the United States (the forty-eight contiguous states, that is), three grids are in operation. Most of Canada and parts of Mexico are also included in those grids. For now, think of a grid as a single large machine. In fact, the electric grids are the largest machines made by humankind. (In saying this, I offer my apologies to the people who built the Large Hadron Collider, or LHC, in Europe, which is often mentioned as the largest machine ever made by humankind. I have heard or read that statement multiple times, but I must disagree. The LHC is 27 kilometers in circumference. The eastern grid is interconnected from Canada to Florida, moving west to the Rocky Mountains.) The three grids in North America cover a huge area (see diagram 1-1), serve over 330 million people, and have more than two hundred thousand miles of transmission lines (these are just the high-voltage lines; distribution lines are far more numerous).

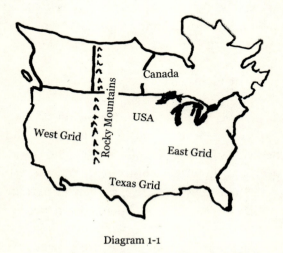

Diagram 1-1

The Three Grids of North America

Starting with a Small Electric System

The best way to understand the grid is to start with a small electric system and build up from there. Indeed, this is how the grid came to be. I will start by considering a small portable generator serving some load in my house. Say a hurricane has knocked my lights out and I, wisely, own a generator since my wife demanded that I buy one. I put gas into the machine and start it up. It runs. I then use extension cords to hook up some lights and plug in our TV.

I notice that the generator runs a little harder every time I add a new device to it. This is because there is a neat controller built into the generator that governs the speed at which the generator runs. This controller is called a *governor*. Every time I connect another device (load) to the extension cords plugged into the generator, the generator slows down a little and then very quickly boosts power in order to regain the desired speed. Without the governor, the greater the load on the machine, the slower it will go. What happens is that the governor senses that the speed has slowed down a small amount, so it boosts the fuel going into the machine to make it regain speed in order to run at the desired speed (in the United States, we like to run our generators so that the frequency of the system is nearly sixty cycles per second).

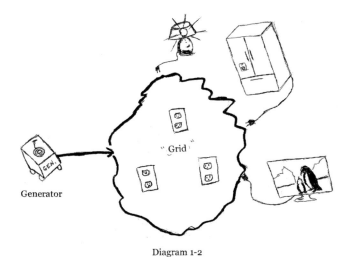

Diagram 1-2

A Grid in Your Home

The example above, diagram 1-2, presents a very simplified version of a grid. Think of the extension cords as the grid. The grid connects the generator to the load. The control is really very simple in that the governor senses speed and attempts to maintain as close to a constant speed as it can.

With my generator, I may also have control of the voltage output, so I set the voltage for the US standard of around 120 volts. As the owner of the machine, I control the amount of load put on the generator. But if I put more load on the generator than it is rated for, what will happen? Without the generator's having some form of protection against overloading, it is probable that the overload will eventually cause the generator to fail. However, there should be some protection built into the generator that will shut it down before it burns up.

Without going into an extensive review of the history of electric systems, I will say that the example above is not too different from how electric systems started. Cables were strung up on poles to deliver the electricity to lights and streetcars within a city. The size of the generator, or its capability to produce power, was designed to match the amount of lighting load expected to be connected to the cables. When the single generator ran out of fuel or broke down for some reason, all the lights connected to it went out.

The Difference between Power and Energy

I will digress for just a bit here and mention the difference between power and energy. I'll use the example of the small portable generator that is running a few devices while the electric service is out. The generator has a power rating. For now I'll say that it is rated for 2000 watts. This is the instantaneous power that the machine can produce. In this example, I have connected 1000 watts of load to the generator. This would be equivalent to ten 100-watt lightbulbs. Therefore, the generator is producing 1000 watts of *power*. If this continues for one hour, the generator has produced 1000 watts of power for an hour, or 1000 watt-hours of *energy*.

When power is produced for a segment of time, the product of power output multiplied by the amount of time is the energy. Actually, 1000 watts is the same as 1 kilowatt (*kilo* means "one thousand"). Likewise,

1000 watt-hours is the same as 1 kilowatt-hour. The utility company typically charges for energy based on kilowatt-hours, or kWh. One thousand kilowatt-hours is a megawatt-hour, or Mwh.

In the business, we often refer to the power flow on a transmission line. We might say, "There are 175 megawatts on that line." This terminology refers to an instantaneous reading, so it is about power. I've seen lots of technical people get the difference between power and energy mixed up, so if this is confusing to you, don't despair.

The Growth of Small Electric Systems

As the demand for electricity grew in the late nineteenth century, it became necessary to add generators. Therefore, multiple generators were interconnected and thereafter served increasing amounts of load. The controls for these small systems were designed to provide the energy as reliably as possible and at the lowest fuel cost.

Imagine a system with two generators running at their rated output, connected together, and serving loads also equal to the combined generator output (see diagram 1-3). One thing that is always important to remember with these connected systems is that the generator's power output and the power delivered to the load are equal (for now, let's ignore losses). In this example, the generators are rated at 10 kilowatts of power output each. Also in this example, the load connected to the system is 20 kilowatts. This would be equivalent to about two hundred 100-watt lightbulbs. What happens when one of the generators suddenly breaks down? The load on the single generator that is running is twice the rating of the generator. In this simple example, there are a few possible outcomes, where

- the overloaded generator will burn up while trying to serve this large load;
- a segment of the load equal to 10 kilowatts will be removed quickly; or
- the overloaded generator will shut down by protective devices that sense the problem, and therefore none of the load will be served.

My bet is that the entire system will black out.

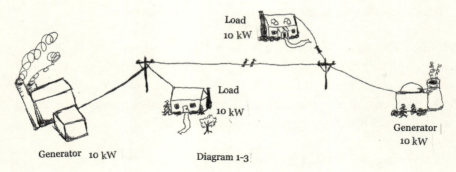

Diagram 1-3
An Early Electric System

In the early days of electricity, the general public was probably very excited about the idea of having electricity. But after the honeymoon period was over, the constant on-and-off of the system came to be a nuisance. The customers demanded a higher level of reliability. Is this much different from what happens today? It was clearly evident that these small electric power systems were more reliable if they had multiple generators running simultaneously. In fact, if the system could be configured so that the loss of the largest generator did not result in a complete loss of the system, then reliability would be greatly enhanced.

To carry the example above a little farther, imagine that there are three generators connected together with the load (see diagram 1-4). Each generator is capable of 10 kilowatts, and the load is, again, 20 kilowatts. So in this example, each generator is running at two-thirds of its rating of 10 kilowatts, or at about 6.67 kilowatts each. Now if one of the generators is shut down, the other two generators can ramp up to carry the entire load. In this example, the load is served continuously. Thus, as this simple example shows, the growth of the electric system had the added benefit of improved reliability. The more robust the system, the more it could withstand the loss of generators or, for that matter, the loss of other components in the delivery system.

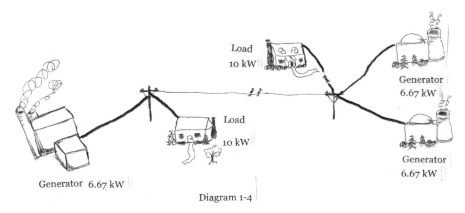

Diagram 1-4

A Slightly Improved System

The Next Step in Grid Development: Interconnecting

The next step in the evolution of the grid was the interconnection of systems. Rather than buying additional generators in order to build the robust system needed for reliable delivery, the companies that owned these electric systems began to interconnect with similar companies. That is, they simply tied the systems together with a transmission line (i.e., a tie-line). (See diagram 1-5.) Of course, there were some complex contractual arrangements to work out, but the benefits justified undertaking that work. The concept was simply that robustness of the system improved reliability. As these electrical systems grew, reliability increased, even though there may have been multiple owners of the equipment. This process went on for some time, starting in the big cities and eventually getting to the point where some of the tie-lines went through rural areas to connect cities.

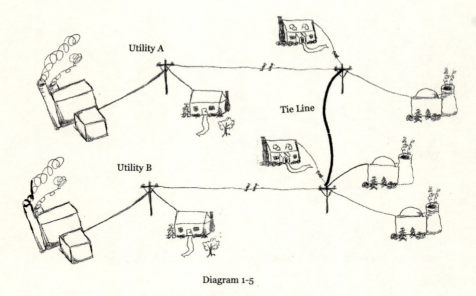

Diagram 1-5

Early Electric System Tie-Lines

Another benefit of interconnections is the potential for reduced costs. Those reduced costs can be found in capital investment as well as in fuel costs. For example, the people at utility A decide that load demand is growing rapidly and they want to build a generator that is big enough for the load ten years out. But since they don't really need all that capability for the next several years, they work out a deal with interconnected utility B to sell a percentage of the output of its new generator to them for the next five years. This is then a good deal for both companies. Utility A gets some return on investment in the early years before their load requires all the output, and utility B gets the generation output they need for their load for a few years without building a new generator.

There is also a great potential in this situation for reduced fuel cost as well as for reduced capital investment. With multiple generators available to serve the load, the lowest-cost generators could be run more than the higher-cost generators. With a simple fuel-cost-sharing agreement between the two utilities, a lower-cost delivery is the result.

But just like most things in life, interconnecting electric systems is a double-edged sword. It certainly helps a system to be more reliable, lowers capital investment, and allows for transfers of energy to save costs. But there is, as there always has been, the possibility of having an interconnected neighbor do something that negatively impacts one's own system (the northeast blackout of 2003 was caused by one or two companies, but it wound up affecting many others that were interconnected). The pros and cons of interconnection have been debated by utilities over the years, but overall the answer is clear: interconnecting is the best way to go.

> Don't confuse the design of radial and network systems on the grid with the concept of series and parallel circuits. Series and parallel circuits are important for design of products, but for the purposes of the grid delivery system all loads are in parallel, whether on a radial feed or a network feed.

Connecting the Grid in a Network

I want to step back here and explain another complexity of the electric system. That complexity is the difference between a radial system and a networked system. In a radial system, the electricity can only go one direction: toward the load. The power leaves a substation and heads down a power line, very often overhead, to the load (such as a house). (See diagram 1-6.)

This is how most of the residential customers receive power. If that one radial line is lost for any reason, then the load is not served (i.e., the lights are out). Most of the electric system is built on this radial principle. In fact, the electric distribution in a house or an apartment is radial. The line comes through the meter into a box, which has a circuit breaker or fuse protecting each line from the power box, such as the circuit that connects all the receptacles in a room. If a circuit is overloaded (that is, if too much stuff is connected to the receptacles), then the fuse blows or the circuit breaker trips. Then all devices on that circuit are out until something is done about the blown fuse or tripped circuit breaker.

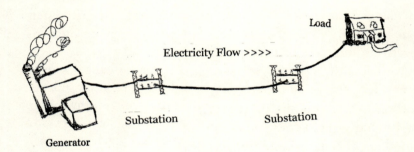

Diagram 1-6
Radial Electric System

However, there is a more reliable and, of course, more complicated way to design the interconnections in the grid. This alternative method of interconnecting is called a *network system*, where the power can actually have several alternate paths to follow to get to another point. The transmission lines in a network system are interconnected so that they form a grid of interconnections. (Take a look at diagram 1-7; notice that the substations in the middle of the grid have multiple sources.) An analogy for this would be the highway system. In most cases there are alternate paths to take to get from point A to point B. They may not all be equal from the standpoint of distance or traffic, but they will get a driver to his or her destination, as the road systems in this country are generally a network, or grid.

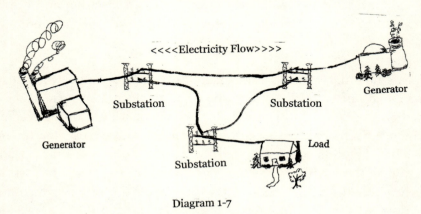

Diagram 1-7
Network Electric System

At some point in the evolution of the electric power system, the delivery of power was split into two levels: the distribution system, which actually delivers the power to the loads and is usually radial, and the transmission system (the grid), which is used to take large amounts of power from multiple generators and then deliver it to the distribution systems. This grid is connected as a network. The dividing line between transmission and distribution is somewhat arbitrary, and yet the function of each is different. It is important that the reader understand the concept of each. A network connection can be more reliable than a radial connection, but the network is far more complicated. Much of this book will delve into the good and the bad characteristics of a network-connected grid.

One additional point is how the term *grid* is used very loosely in the popular press and on the Internet. I have seen the term applied to the entire electric system, all the way to describing the delivery to someone's house. As we shall see in chapter 8, many of the applications that fall under the term *smart grid* are actually used at the distribution level. My definition of the grid, however, does not include the distribution system, which is mostly radial. The term *grid* is used to describe the network of interconnections that comprise the electric transmission system. I am not arguing that there is a right and a wrong terminology. For purposes of clarity, however, I will be describing how the transmission system works. An easy way of differentiating the two is by the voltage level. Let's say that anything below 69,000 volts is in the distribution system and that anything 69,000 volts and above is in the transmission grid.

Alternating Current vs. Direct Current

The term *frequency* will be used a lot, so I want to make sure the concept is clear. *Frequency* is the term used to measure a repeating event. I will use as an example the heartbeat, as everyone should be familiar with this biological process. The heart goes through a cycle (called a *cardiac cycle*) about sixty times per minute for the average person at rest. Each beat is a cycle as the heart goes through the major steps involved in blood flow. Therefore, the heartbeat frequency is sixty cycles per minute. Another example of frequency is the radio signal. On the AM dial, there may be a signal with a frequency of 1000 kilohertz per second. *Hertz* is a shorthand term for "cycles per second," so in this example the frequency of the signal oscillates at 1000 × 1000 (kilo = 1000) cycles per second, depending on which station is tuned in. That would be shown as 1000 KHz on the dial.

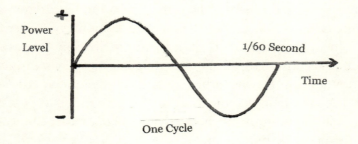

Diagram 1-8
Alternating Current

The grids that I have been talking about are based on an alternating current system. That is, the voltage and current (and hence power, which is simply voltage × current) are produced as a sine wave (see diagram 1-8) that alternates at sixty cycles per second (Hz). Alternating current is the preferred method of delivery of power in this country. Another form of electric power delivery is direct current. Direct current is simply like the output of a battery, with a constant voltage output across a positive and a negative terminal. There is no frequency related to direct current since there is no oscillation of the output.

There was a big debate in the early days of electric development (from the 1870s into the twentieth century) about whether direct current or alternating current would be the best way to deliver power to the customer. This debate included some familiar names: Thomas Edison was a direct current proponent; George Westinghouse and Nikola Tesla were both alternating current proponents. Direct current (DC) systems were actually first on the scene, in New York City, and Edison had some patents on DC devices. The main problem with DC, especially evident at the time, was that there was no easy way to change the voltage level in order to transmit the power. So whatever voltage was produced at the generator was put on the overhead cables and delivered to the load. Unfortunately, since the voltage would drop as it went farther from the source, the result of resistance in the cables, the loads could not be very far away from the generator. This is a problem because the devices that use the electricity are designed to function within a small range of voltage; they won't operate correctly if the voltage is too low.

Alternating current systems can use a device called a *transformer* that simply converts voltage from one level to another, determined by the number of turns in the coils. Transformers allowed the voltage produced by the generators to be raised to higher levels so they could be transmitted long distances and then be reduced to the level that the customer needed. The transformer made it possible for the system to evolve as we know it today, with generators many miles from the loads in most cases. Suffice it to say that alternating current won the debate. Systems that delivered direct current to loads designed for direct current were mostly phased out in the twentieth century, although it is possible that some very small DC systems still exist that I haven't heard about.

Building an Interconnected System

With the information I have presented so far, the outcome for the grid should be easy to predict. All the little companies interconnected, the load continued to grow, and interconnected companies then formed interconnections with other interconnected companies until the entire electric system was interconnected. Well, almost all. I did mention there are three grids, or interconnections, in the United States. Why not just one? Well, the eastern and the western grids evolved a lot like I've discussed, but the two were not interconnected. The Rocky Mountains were an effective barrier to an interconnection, and there really wasn't a lot of additional value in connecting these two systems together. I read somewhere that attempts were made to tie the eastern and western grids together back in the 1950s, but the engineers could not find a way to tie the systems together permanently. I don't know if that story is true. Anyway, today the east and the west are connected by some direct current links only.

I said that the eastern and western grids are not interconnected, but then I may have confused the issue by saying that there are some direct current links. To erase confusion on that point, I will introduce another important point to the discussion. The most important concept in the definition of a grid is the matter of synchronicity. For a system to be considered one interconnected grid, all the generators providing power

> For a system to be considered one interconnected grid, all the generators providing power must be synchronous.

must be synchronous. What this means is that each generator's sine wave output is aligned with the sine waves of all other generators. Diagram 1-8 shows the wave form that alternating current follows. All the generators supplying power are following the same wave form and, consequently, the same frequency for the delivery of the power. The frequency is imposed by the physics of the entire grid. In other words, all the generators on a single grid are in lockstep.

As an example, think of a choreographed dance routine performed by a large troupe of professional dancers. The music that plays sets the speed of the dance, and no single dancer can change that fact. If any dancer gets out of step, the only way the routine can succeed is if all the other dancers continue in step, meanwhile hoping that the one dancer out of step either gets out of the way or gets back in step. (But the troupe probably gets pretty irritated at the dancer who lost the imposed rhythm.)

I have mentioned DC ties between grids, but I have not described what they are. These DC ties are pretty neat. The way DC ties work is that the alternating current from one grid is converted to direct current, using some pretty high-tech equipment, and is then connected to one end of the DC tie-line. The DC tie-line may be hundreds of miles long, and it is usually at a very high voltage. At the other end of the DC tie-line, the same type of high-tech equipment is used to convert the DC back to alternating current where it is connected to the other grid (see diagram 1-9).

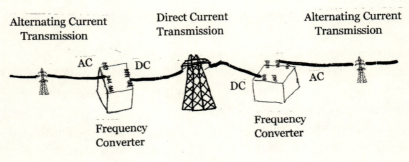

Diagram 1-9

Direct Current Tie-Lines between Grids

A DC tie between two grids would not impose the synchronicity requirement to the grids on each end. Therefore the grids on each side of the DC line could be out of step, or asynchronous, and it would not matter a bit to the DC tie. Also, power can be controlled to flow in either direction. It can also be controlled to allow only as much power through the tie-lines as desired. Each grid is allowed to float along at whatever frequency it happens to be running, asynchronously with the other grids. Another interesting characteristic of the DC tie is that problems in one grid are not passed along to the other.

It turns out that when transmitting large amounts of power long distances, direct current is preferred. This is because losses are reduced by the use of high-voltage DC. (Losses are a by-product of delivering energy through the electric system. There is no practical way to avoid them completely. A rough guess of the amount of energy that is "lost" from the generator output to the load is 7 percent.) It is for these reasons and several others that DC ties are used between all three of the grids in the United States.

Texas: State vs. Federal Jurisdiction

Moving on to Texas, the interconnection there is considered a separate grid also. However, this grid is connected to the east and west grids with DC ties. Why isn't the Texas grid integrated with the east or the west grid? The answer is found in the US Constitution. Where the Constitution gives states the jurisdiction to govern all intrastate commerce, the federal government takes jurisdiction over interstate commerce. Since a synchronous grid that is interconnected, as discussed previously, crosses state lines, transactions that take place on that grid are considered federal jurisdiction. Since the Texas grid is entirely within the state of Texas (but I will note, just to complicate things, that some parts of Texas are not in the Texas grid), jurisdiction of commerce on that grid is not given to the federal government. This is what the early Texas developers of their grid wanted to accomplish, and they succeeded. The flows over the DC tie-lines that cross state lines are under federal jurisdiction. However, without getting into any further legal explanations, I will say that there are certain rules issued by the federal government that still apply to the Texas grid. Most important to our discussion are the rules related to reliable operations.

An Analogy for an Interconnected Grid

So far, we have a large (networked) interconnection of transmission lines and generators, connected to substations that distribute power to the customers (radially for the most part—the main exception is in big cities). Power is put on the grid by the generators, where it is comingled with all the power generated by all other generators and is then delivered to the substations. A way to understand this is to think of a barrel half full of water with several pipes bringing water to the top of the barrel and with several pipes at the bottom of the barrel taking water out (see diagram 1-10).

As long as the flow of water into the barrel and the flow out of the barrel are equal, the water level in the barrel stays the same. If there is more flow into the barrel than out of the barrel, then the barrel slowly fills up. Conversely, if more water is draining out than is replaced by the pipes at the top, then the barrel slowly empties. Of course, the pipes supplying water at the top represent generators; the pipes at the bottom of the barrel represent load. The level of the water in the barrel is the analogy for the frequency of the system. Similar to the barrel, if there is more generation-supplied power than the customers' load demand, then the frequency of the grid actually increases, much like the level of water in the barrel would increase.

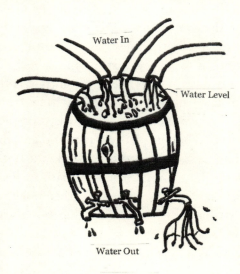

Diagram 1-10
Water Barrel Analogy for the Grid

To make the analogy work, I'll assume that the amount of water leaving the barrel is proportional to the level of water. That is to say that an added level of water would put more pressure on the pipes at the bottom so that more water would flow. And of course, conversely, a lower water level would mean less water flowing out. So now if more water flows in, the water level increases and more water flows out until a new equilibrium is reached.

The electric grid acts similarly. When generation increases slightly, frequency increases slightly—and as this happens, the load increases as well and a new equilibrium is reached, at a slightly higher frequency. Imagine for a minute an electric motor turning a pump that is pumping water. The motor's speed is proportional to the frequency on the grid that is supplying power to it. As the frequency of the grid is increased, the motor spins a little faster. The pump also spins faster, which results in more water being pumped. Pumping this additional water requires more instantaneous power.

So the increase in power output by the generators is used by the load at a slightly higher frequency. Take this example and multiply it by millions of devices, and it should be evident why the grid would simply find a new equilibrium speed where the input power and output power are matched. Remember that the generated power and the power delivered to the load are always equal (again ignoring losses). The grid does not store power (or energy).

Grid Evolution Was Not Random

Essentially, all three of the grids in the United States are similar in most respects except for size (the eastern one is far larger). There are a large number of transmission-owning companies that have developed the grid as we know it today. In fact, there are over a hundred companies that own parts of the grid in the eastern interconnection. The grid is very highly interconnected, but it was not just randomly developed. The companies that built the transmission

> There are over a hundred companies that own parts of the grid in the eastern interconnection.

grid had engineers who study how power flows on the grid for a large number of different situations.

These studies would include analyses considering load growth for ten- to fifteen-year horizons, as well as planned and probable generation additions over that time frame. Remember that the networked interconnection offers many paths for the electricity to flow. The studies have to take into account that the electricity will flow on the path of least resistance (a law of physics for electricity). To complicate matters even more, the path the electricity flows through on the grid varies based on all the grid interconnections at that point in time.

The engineers' analyses would also consider unplanned events such as the sudden loss of generators and transmission lines. The engineers would try to design the grid in such a way that the system would not black out for these kinds of events. Studies are also performed jointly by companies that serve a large region to make sure that the impact of these events across multiple companies could be contained without a blackout. The point here is that the grid evolved and grew over the years, but not without a great deal of planning.

Ownership of the Grid

This is a good place to discuss the ownership of the grid itself. Up to now, I've mentioned that there were many companies involved in building the grid, but I haven't gone into any detail about the grid's ownership. I have to say that I, after working in this business for thirty-eight years, still find it to be a complicated puzzle. Not only are there multiple types of organizations involved in the grid, but also there has been constant change in those organizations since the deregulation of the wholesale market in the mid-1990s.

The largest piece of the puzzle is the investor-owned utilities (IOUs). These utilities are considered to be vertically integrated, as they own generation, distribution, and transmission, providing complete service to their customers, including billing. Each has been granted a certificated territory to serve, wherein they have the exclusive right to sell retail electricity in those areas, but in return they must serve the entire load demand in their area (at a reasonable price). The IOUs typically own, maintain, and operate the grid in their area, with some exceptions for joint ownership of tie-lines and other special arrangements. Examples of IOUs are Duke Power, Florida Power and Light, Ameren, and Dominion Virginia Power, among many others. To provide some perspective on this, IOUs serve approximately two-thirds of the load in the United States.

The next big piece of the puzzle is the Tennessee Valley Authority, established by the TVA Act of 1933, which operates as a federal corporation. The TVA was established to provide electricity and to assist in economic development in the Tennessee Valley. The TVA operates in seven states, selling power to distribution companies and industrial companies but not to residential customers directly. TVA owns, maintains, and operates the grid in its area much like the IOUs do in theirs. TVA is subject to the reliability standards for the grid that are discussed in chapter 2, as they are subject to the Federal Energy Regulatory Commission's (FERC) jurisdiction over the reliability of the grid. FERC does not have jurisdiction over TVA's electricity sales since the statutes that define FERC's role do not include that authority.

And if this isn't complicated enough yet, there is still another important federal player called the *power marketing administrations* (PMAs). There are four PMAs in the United States that are tasked with selling the output of the federally built hydro generators to recover the government's investment. Included in this group are Bonneville Power Administration, Southeastern Power Administration, Southwestern Power Administration, and Western Area Power Administration. Several of the PMAs own transmission, Bonneville especially, and are big players in the puzzle we call the grid.

Another important piece of the puzzle is the cooperatives. These organizations were formed by the Rural Electrification Act of 1936. They are nonprofit and member-owned, but in many ways they are similar to the IOUs, at least in the way they operate. There are many, such as Associated Electric Cooperative, Inc., that own transmission and generation as well as distribution, making them vertically integrated, much like the IOUs mentioned before. Many others are distribution-only, although many of these organizations also own generation. Very often this ownership is joint with IOUs.

Another player in the generation, distribution, and sale of electricity is the municipalities. There are many municipalities that own and operate their own distribution system. There are also many that own and operate generation. As far as the grid is concerned, there are many that own transmission, such as Atlantic Municipal Utilities. I should point out here that the concept of transmission is based on a voltage level only. Although the equipment owned may be for purposes of transmission by that definition, it may not be networked into the grid; instead it is radial.

Still another type of player is the transmission-only companies that have formed fairly recently, such as American Transmission Company (ATC) LLC, formed in 2001. These companies are significant in size on the grid (ATC has over nine thousand miles of transmission lines, with over 12,000 Mw of peak load served in 2012) and are an important element in maintaining the reliability of the grid.

Another Technical Concept: Voltage

One technical concept that I have yet to describe, but one that is important to understand, is the voltage of the grid. Voltage is the equivalent to water pressure in a hose. The more pressure in the hose, the more water comes out, and the farther it travels when it does. Without the water pressure, nothing happens. In an electric circuit, voltage is analogous to water pressure. Voltage is what actually forces the electricity to flow to lightbulbs or phone chargers. In residences in the United States, voltage is around 120 volts, with an option of going to 240 volts for some devices such as clothes dryers. Remember too that this is alternating at 60 Hz.

It may seem at this point that the grid is all interconnected at one voltage level, but that would be too simple. In reality, the grid was first built at lower voltages. As the system grew, it required higher voltage levels in order to efficiently move electricity from the generators to the substations that serve customers. Different voltage levels are interconnected by devices called *transformers* that can move power through them from a low voltage to a high voltage or vice versa. The transformers allow the interconnection to work together across many different voltage levels (see diagram 1-11). Just as a data point: the highest voltage today in the eastern grid is 765,000 volts.

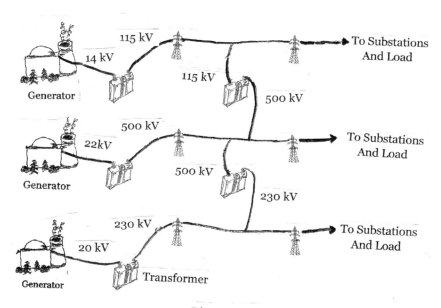

Diagram 1-11

Transformers Allow the Grid to Connect Together

The Three Electrical Phases

There is another technical characteristic I will mention in order to clarify the physical construction of the grid. The power delivery in this country is, as I mentioned earlier, at about 60 Hz. However, this power is transmitted in three phases. By this I mean that the sine wave in diagram 1-4 is actually only

The power conductors for each phase may be more than one huge wire. In fact, some of the larger lines use a bundle of conductors for each phase. So there may be a bundle of four individual stranded cables to carry the power on one phase.

one phase, and the other two phases are displaced in time by 120 degrees. What do I mean by 120 degrees? One complete cycle of a sine wave is considered to be 360 degrees, so 120 degrees is one-third of a cycle (see diagram 1-12). It's kind of like singing a round of the nursery rhyme "Row, Row, Row Your Boat" with three separate groups starting at different times.

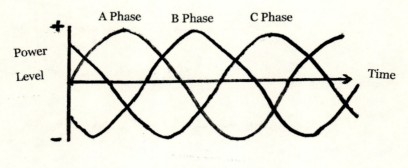

Diagram 1-12

The Three Phases of Alternating Current

The construction of large transmission lines includes three separate power conductors for each line, or to be precise, one for each phase (see diagram 1-13). The transmission line is represented in most of the sketches in this book as a single line. However, there are actually three phases carrying power in this line. It is three-phase because way back when the electric system was first developed, the engineers and accountants realized that three-phase generation and delivery is more efficient (i.e., cheaper) than is generation using one, two, four, or more phases. Distribution lines can be three phases, although many times in some small neighborhoods there may be only one phase. And of course DC lines have only two conductors.

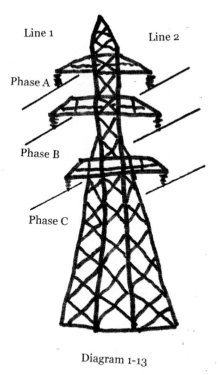

Diagram 1-13
Three-Phase Transmission Line

How Electricity Flows through the Grid to a Home

To illustrate how the electricity flows through the grid, I will follow the path from a generator to the load, such as the electric load in a home or an apartment. For technical reasons, electricity is generated at a lower voltage than is carried on the local grid, so after the power is generated, a transformer is used to step up the voltage to a higher level to match the voltage of the local grid. This electricity is then comingled with the output of other generators in a big soup on the grid (remember the barrel analogy). Then the electricity flows from the soup into a substation, where it goes through another transformer, this time to lower the voltage to a level suitable to distribute the electricity to residential areas. From the distribution line there is another, smaller transformer that lowers the voltage again to the voltage actually used in a home or an apartment.

William L. Thompson

The electricity that finds its way to the load in a house is from a multitude of generators. Lots of people have asked me over the years, "Which generator is serving my house?" There is no correct answer to this question, as the grid is constantly changing, which impacts the flow of electricity. Perhaps the best answer is to provide an estimate of the percentage of energy of each type of fuel source that all customers share for the local utility. This might come out as something like 35 percent nuclear, 30 percent coal, 25 percent natural gas, and 10 percent hydro and other renewables such as wind and solar. Of course, the generators that are closest to a particular house are probably providing a higher proportion than these averages, but to know how much at any given time requires a complex analysis.

When I talk about generation, I may be referring to a plant or a station, such as a nuclear plant or station, or I may be referring to the individual units at that plant, which essentially can operate independently. There are usually multiple units at any generation plant or station, with some shared facilities, such as the coal pile at a coal plant. I use this terminology often, so I thought I should make it clear.

Physics has shown that electricity flows from the higher voltage source back to the zero-voltage terminal at the source. Electricity will always "find" the easiest path to get back to the zero-voltage terminal. This path is often called the *path of least resistance*. So even though the output of the generator nearest to a home is comingled with the output of every other generator on the grid, the electricity from that machine probably will not flow very far before it finds its path, through the load, back to its source. The reason for this is that the transmission lines that make up the grid have a resistance to the flow of electricity. The farther electricity has to travel, the more resistance it must overcome. It is very unlikely that a generator in Michigan would be serving load in Florida, although such a thing could happen—since the two states are interconnected by the grid—if the transmission path had a low enough resistance. The flow of electricity from the generator to the home happens virtually instantaneously, as electricity flows at the speed of light. In fact, the electricity that I am using in order to type these words on my computer was generated essentially at the same time as my typing. As I mentioned before, there is no storing of electricity on the grid; whatever is generated returns to the source at nearly the same instant it is generated.

Bad Things Can Happen to the Grid

When all is working properly, the electric grid is a fine machine that serves us all well. However, it is constantly challenged by factors that could cause major problems. Electrical devices fail, usually unexpectedly. And there are a lot of electrical devices that make up the grid. When something fails, the grid has to be able to isolate the problem and keep on doing its function. There are also a lot of humans who do things every day that can affect the grid. Human error can challenge the grid's performance as much as equipment failures can. Most of the problems that occur on the grid are contained; in fact, no load is lost for most of these events thanks to the nature of networking, as discussed above. Occasionally there are events on the grid that cause outages to large areas with many customers. These events might include tornadoes, hurricanes, and major equipment failures. The worst-case event that electric utility folks worry about is the loss of the entire grid, a blackout.

Previously I used an analogy likening the grid to a barrel of water with a nearly constant level. Imagine for a minute that on account of some problem, like a loss of water source, the barrel has completely emptied, that all the pipes are there but no water is flowing. This is the analogy for an electric grid blackout. The challenge now is to get back to the point where the barrel is half full of water, with the amount of water flowing in matching the amount of water flowing out. I imagine that it would take a step-by-step process with the barrel, first filling it halfway and then carefully opening valves at the bottom and at the top to get back to the point where the inflow matches the outflow. But here the analogy with an electric system breaks down. Since electricity is not stored, generation and load must match instantaneously all the time. Recovering an entire grid from a total blackout is a very difficult undertaking, since the generation and load must be matched constantly.

When really bad things happen to the grid, the grid may reach a point where it collapses or blacks out. This can happen to parts of the grid, as happened in the northeast blackout of 2003, or, in even worse cases, for the entire grid. In 2003, there were over fifty million people without power, many of them having no power for several days. However, this was only a partial blackout of the eastern grid. After things calmed down, the remaining portions of the grid were available to help restore power to the locations where it had been lost. The point I am trying to make here is that the grid allows for an accustomed standard of living yet is not immune to failure.

Stories from a Control Room

Over the years that I worked in the System Operations Center at an electric power company, I performed hundreds of tours for people with an interest in how the grid worked. I enjoyed doing the tours because it was exhilarating to see people reach that "aha" moment when they understand what is going on. Usually they had no clue what the grid is before they came for the tour. These people asked me a lot of great questions over the years.

One of the best questions I can remember, after a discussion of the pros and cons of interconnecting on the grid, was, "Why don't you have a switch that disconnects your system from the rest of the grid?" This could certainly be done (although the practice would be frowned upon by other utilities). The system that I worked on for my career has over twenty tie-lines with neighboring utilities. These tie-lines are what help to form the grid. Electricity flows on these tie-lines constantly. The idea is that if the electricity flowing on the tie-lines became so great that the system appeared to be on the verge of collapse, an operator could flip a switch to disconnect all the tie-lines simultaneously and Dominion's system would no longer be part of the overall grid. This might protect Dominion Virginia Power's service area from an impending blackout if we saw it coming soon enough.

My answer was that although that switch might save us in some very rare event, the most probable outcome would be a hastening of the collapse of the system. Indeed, the collapse would be even more likely if all transmission owners installed such a system. And then there is the law of unintended consequences regarding the probable accidental operation of the switch at some point, which could then cause a blackout. It is amazing to me the large number of times these unintended consequences happened. If the separation switch was installed it would have to be tested and maintained over the years to make sure it will work if needed. I suspected that one day when least expected, the switch would be operated inadvertently and cause some serious problems for the operators.

Also, disconnecting from the grid may not be the right thing to do, even in challenging situations. Just prior to the northeast blackout of 2003, the utility at which I worked experienced some strange flows and some very low voltages. After the blackout, though, the remaining grid stabilized quickly. Had we elected to disconnect from the grid, it would have made things worse for us—and potentially for other utilities as well.

As the year 2000 approached, people who work with software warned management people throughout the world that there was an inherent problem. This was the Y2K scare of the late nineties. I have been told that there were software routines designed that used the digits of the year for some calculations. As long as the year was 1999, this worked fine, since "99" was used in the calculation. When "00" would be used, however, the concern was that our computers were going to freeze up, especially if the program divided numerals by 00. Some analysts predicted that the entire electric system would black out.

A lot of time and effort went into fixing our software in the late nineties. But did everything get fixed? How could utility operators know for sure? On the night of December 31, 1999, the control room at Dominion Virginia Power was full of people. Almost everyone in the company was on duty or on call, and it was probably that way throughout the world. Electric utility people were not doing much celebrating that night. But nothing unusual happened, I'm only too happy to say. In the US we had the advantage of knowing that nothing happened in Japan, Australia, and other places where the date changed earlier than it did for us. It seems that the software people did a great job of correcting the software. But I must admit that the anticipation was a bit stressful.

William L. Thompson

A lot of people I gave tours for did not realize that the clocks on the wall, at least those that are plugged into a power outlet, are keeping time using the frequency on the grid. We target operation at 60 Hz, but because of small variances in generation and load, the grid frequency oscillates around 60 Hz all the time. In fact, if the load is consistently greater than the generation, then the frequency stays below 60 Hz. On especially hot days in the summer, those of us in the eastern interconnection would often see this happen. The error in the clock time would be a known quantity since our computers could integrate the frequency error over time and calculate the time lost. One day the clock error was over ten seconds slow as shift change approached. One of the operators mentioned to me, "If this continues, we'll never get off this shift!" Imagine all those kids in summer school watching the clock and wondering why the day seems so long.

Chapter 2

GRID CONTROL

It was one summer in the early 1990s when a group of high school science teachers requested a tour of the grid control center. My group at Dominion was operating at an old control center, a very small single-story room with no viewing area, so we had to take the tour on the operating floor and walk between the operators' desks to give folks an idea what was going on and why (this was a long time before 9/11 and the thought of terrorism). One schoolteacher reached for a control switch handle and asked, "What does this do?" My reaction was close to panic; just the thought that someone might try something like flipping a switch had not crossed my mind. The teacher was just kidding, but that was the last tour we gave on the operating floor.

Introduction to Control of the Grid

How the grid is controlled is a great question that I have been asked many times. It has no easy answer, however. There is no single controlling entity for the grid, and no single entity knows all that is going on at any given moment. In a sense, the grid is a lot like the Internet. To provide some perspective on this concept, I'll mention that there are numerous entities that control different geographic areas of the grid, and even more entities that control other parameters that affect the grid. In the eastern grid, there are at least a hundred entities with some control, and that is just for the transmission portion of the grid. There are others that are controlling generators, and still others that control the distribution system. To a large extent, the grid is controlled by all of these controlling entities that are doing what they are supposed to do, without any direction in real time from a single controlling entity.

But before I delve into answering the question of how the grid is controlled, I think it would be helpful to explain the things that can go wrong on the grid. Having explained these risks, I can then go into the detail about how these risks are mitigated. In approaching risks this way, it should be clear why the grid needs to be controlled. (By the way, the switch for which the schoolteacher reached would have done nothing.)

Why Does the Grid Need to Be Controlled?

The number-one overriding concern for all of those involved in grid control is to prevent a blackout. A blackout is the "eight-hundred-pound gorilla" of bad events. Any discussion of grid control, access rules, good operations, etc., will eventually lead to the basic concept of preventing a blackout. The cost to society of a blackout is immense in terms of dollars, but the cost may be even larger when considering the resulting absence of safety in our way of life. Although the event of a total blackout is studied and tested on simulators around the country, there is no way of knowing how long it would take to fully recover from a complete grid blackout. I say this because the loss of power could affect many of society's infrastructures that are depended upon for recovery. An example is the need to get service people out to substations when the traffic lights are not working and traffic is backed up for miles.

There are a number of factors that if left uncorrected can lead to a grid collapse. For example, in order to have a healthy frequency on the grid, generation and load need to be balanced or at least kept reasonably close to a balance. In addition, problems caused by storms or equipment failure must be contained. Below is an outline of the issues that I believe have the highest potential of leading to a blackout if left uncontrolled. After describing these issues in some detail, I will then describe how experts who work in this business are able to minimize the risk posed to the grid.

Following are the key issues I will discuss:

- reactive power
- generation and load balance
- storms or equipment-failure events
- generation loss
- system design / cascading events
- stability
- markets
- solar magnetic disturbances

> These are the main issues that challenge the integrity of the grid.

Reactive Power

As I mentioned before, one event that we know can bring down a grid is a voltage collapse. A voltage collapse can occur in large parts of the grid where the load is greater than the generation available in that area. The problem is that when electric power is delivered over long distances (for example, twenty-five to more than fifty miles), one property of alternating current becomes very important. This property is called *reactive power*. Reactive power is one of those complex technical concepts that I said I wouldn't get into, but I must explain it, as this concept is crucial to the delivery of electricity. In order to deliver real power in our alternating current grid, there must be reactive power. Reactive power does not provide real power or energy, but it is a necessary ingredient for the grid to deliver real energy. Think of it as the oil in an automobile engine. The oil doesn't provide any energy, but without it the car will not go very far. Another analogy that I have heard many times is that reactive power is like the froth at the top of a glass of beer; it doesn't amount to much, but who wants a flat beer?

Reactive power can be produced by generators or by devices called *capacitors* that utilities install on the transmission system. Unfortunately, unlike real power, reactive power cannot be transmitted very far. The impact of this reactive-power requirement is that only so much real power can be imported into an area (think of this area as a large city, such as Cleveland). In effect, the area in question would need to have its own reactive-power sources in order to maintain the delivery of real power. What this means is that the overall grid can be balanced with frequency at a very comfortable 60 Hz, but the area without adequate reactive resources (such as nearby generators or capacitors) can be at great risk of a voltage collapse. This is one issue that makes the control and operation of the grid very complex and challenging.

Generation and Load Balance

Generation output and the load need to be pretty nearly in balance (that is, with a grid frequency close to 60 Hz) for the grid to be reliable. I'll try to explain by providing the following ridiculous example. Say the grid is doing its thing, loads are being served as needed, generation and load are matched, and frequency is 60 Hz, so all the grid operators just decide to take the day off. It's Super Bowl Sunday, however. As game time approaches, millions of people are switching on their televisions. Therefore, the load on the grid is increasing, slightly at first, but more as kickoff time approaches. What is happening on the grid? Since generators had not been told to increase their output while the load was steadily increasing, the frequency that was 60 Hz has now declined to 59 Hz. Since the generation supplied and the load served always match, as described in chapter 1 (remember, the grid does not store energy), the system frequency declines to a new equilibrium level.

Now the frequency decline I use as an example may not sound like much of a frequency drop, but as the frequency continues to decline, the grid loses its ability to withstand other events, such as generators failing. When a generator fails, it is removed from the grid ("tripped"), either by the action of an operator or automatically. (The reverse of this is never true for significant-sized generators—they don't automatically come on to the grid. Starting up a large generator typically takes hours or days.) Tripping a generator creates a situation on the grid wherein the load is the same as it was but the generation serving that load has decreased; thus the frequency must reduce to a lower equilibrium level. Low-frequency energy delivery is not good for equipment. Without going into detail, I'll just say that the generating equipment and the motors in homes and in factories are at risk of other problems if this situation continues.

There are automatic systems in place to remove load as the frequency reaches low levels (underfrequency load shedding), but this is not a good way to "control" the grid. Many generators are also designed to trip at a low frequency in order to protect the equipment. Personally, I wouldn't want to rely on these systems, as there is no guarantee that they wouldn't cause other problems (such as having generators trip prior to the removal of load, or the automatic removal of load, causing fluctuations that result in stability problems). I'm glad to say that I never had to use the underfrequency-load-shedding equipment.

One more thing about the frequency on the grid: as mentioned above, the frequency on an interconnected grid is the same everywhere during stable conditions. In the eastern grid, the frequency in Maine, Florida, and Michigan is the same. This is true even if there is more generation in Michigan than the load in Michigan. Since the system is all interconnected with networked transmission lines, the extra generation in Michigan will flow to loads in other locations. I've been talking about the system slowing down when generation is reduced, but I haven't given any perspective on the relationship. There is a relationship of generation and/or load to frequency. This relationship varies all the time, depending on the "size" of the grid at any point in time. The size of the grid is the amount of generation/load online at that instant. A ballpark number for this relationship is as follows: 4000 megawatts of load is equivalent to one-tenth of a hertz (0.1 Hz) for the eastern interconnection. In other words, in the example above, where the frequency has declined to 59 Hz, there has been an increase of load of 40,000 megawatts across the grid without any increase in generation.

Storms or Equipment-Failure Events

Any discussion of control of the grid must address the fact of storm or equipment-failure events. I will start with the matter of short-circuits. As mentioned before, a law of physics says that the current that flows will always find the path of least resistance to return to its source. Usually, that path will be through the devices and equipment (loads) that customers want to run, such as televisions and air conditioners. Occasionally, because of a fallen tree leaning on a power line or a broken power line lying on the ground, the current doesn't go through load devices but, instead, goes directly to the ground from the conductor, and the ground provides a path back to the source. When this happens, it is called a *short-circuit* or a *fault*. For the high-voltage transmission system under examination, an electric arc will ensue, very similar to one produced by an arc welder, except that this arc, with the power of the grid behind it, is a lot more powerful.

Remember that the grid is interconnected in a way that allows for the output of multiple generators to take multiple paths to reach a load. What happens is that all of these sources of power are instantaneously affected by the easy path through the ground back to the source, to varying degrees based on their respective distances from the fault. The current that flows on the transmission line that is faulted increases significantly, as electricity in the grid will seek this easiest path to ground. If allowed to continue, the current flowing into the arc, which exceeds the capability of the line conductors, will cause further damage to the transmission line. This damage can propagate to a lot more equipment in the grid if the problem is not remedied.

Not all faults occur in storms, of course. There were many times when something would happen in a substation on a nice sunny day. Often the written report of what happened would come back with the notation "ftr." After seeing this, I had to ask, "What does that mean?"

"Oh," one of the operators with field experience would explain, "that is a furry-tailed rodent that got into a bad place."

"What is a furry-tailed rodent?" I had to ask.

"That's a squirrel. The field folks just got tired of trying to spell it," was the answer.

We often called these events "unplanned events" to differentiate them from the planned outages that we must have on parts of the grid to do maintenance. Whether planned or not, the line in question would no longer be in service. This also creates a problem for the grid's continued operation. To understand this, remember that the grid is networked, which means that there are multiple paths for the current to flow to get from the generators to the loads. In this example, the transmission line that is disconnected is not the only one feeding the local substation, where the distribution system takes over. Now that the line is out of service, the other lines continue to carry the entire load. A simple example would be that if there were three lines, each carrying one-third of the load, and if one of them tripped off, then the other two lines would now each carry half the load.

Unfortunately, it is never quite this simple. Since the current flowing in the grid is always finding the easiest return path to its source, the current in the two remaining lines will depend on many variables, such as length of the lines, proximity of generators, and the actual structure of the grid in the area. In reality, the load is still served, but most likely the current in the two remaining lines will not be evenly divided. These events are quite common. If the grid is not properly designed and operated, these events can cause tremendous difficulties, such as overloading one of the remaining lines.

Generation Loss

Another unplanned event is the loss of a large generator. The term *loss* doesn't mean that the generator can't be found; it just means that the equipment has failed and is disconnected (tripped) from the grid. This happens regularly, no matter how much utility companies maintain equipment or how well they design redundancy in. To put this event in perspective, I dare say that there is at least one generator lost every day on the eastern grid. We've had days at my company where we would lose two or three in one day. In any event, when a generator is lost, there is not a corresponding loss of load. When this happens, the flow of current will change throughout much of the grid and the frequency will decline temporarily throughout the grid. The changes in current flow will be greater in the grid nearer to the lost generator, but the entire grid has to be able to continue to function without any problems after this event. This event can be especially challenging because not only is the real power it was producing lost, but also whatever reactive power it was producing is lost.

System Design / Cascading Events

Carrying on with the concept of why the grid needs to be controlled, I'll point out another area of concern that has to do with the design of the grid. I mentioned that the grid was not randomly built but was carefully planned as it grew. Industry rules establish how robust the grid should be from a design standpoint so that it can withstand some level of unusual events.

Say for a moment that the rules are not in place. A transmission line constructed by one utility could be planned to carry, say, 75 percent of its capacity during normal conditions. This line could be located near the border with another utility. As I've said before, when a transmission line is tripped, the electricity that was flowing through that line will find another path, instantaneously, to the load. In this example the alternate path could be through a neighboring utility's system. If there are no rules in place, then this increased current flow on the neighbor's system could overload several of their lines. These lines then need to be removed from service or else they will be damaged. This could potentially be the beginning of a cascading event that would lead to a blackout of the grid.

Stability

Looking at the failure of the Tacoma Narrows Bridge in the state of Washington in 1940 is a great way to understand the concept of stability. There is a video clip available on the Internet of the bridge actually collapsing (I provide a link under "Suggested Reading"). Basically, the bridge started oscillating when the wind was right at forty-two miles per hour, which resonated with the structure of the span.

Well, electric systems can do the same thing. The issue is called *stability*, and it shows as a generating unit's output oscillating back and forth even though the unit is still connected to the grid. If a generator is disturbed by some event, such as a fault on a transmission line, it will usually oscillate for a short period of time but will still remain synchronous. This oscillation is in the frequency and power output of the machine. The generator's frequency will be a bit faster than the grid for a while, and then it will slow down and be a bit slower than the grid.

Remember earlier I mentioned that the generators must be in lockstep with the grid, that is, they must be synchronous. In this case, which may last for several seconds or even longer, the average frequency output of the generator is the same as the grid, but it is oscillating around that average. As the generator speeds up and slows down, the power of the machine also increases and decreases. If the generator were to go too fast or too slow, it could become asynchronous with the grid, which would inevitably result in the tripping of the unit. The trick is to have the generator oscillations "dampen" or simply smooth out over a short period of time.

Markets

For the purpose of the discussion on control of the grid, it is important to realize the role that the market plays. FERC has moved the electric industry toward implementing wholesale competition. This started in the early 1990s and continues to this day. I want to differentiate the wholesale market from the retail market. Retail electricity is sold for final use, such as to a residence; wholesale indicates electricity sales for resale, such as from a generator to a distribution company. Wholesale deals are made on the grid, and since the grid crosses state boundaries, these deals are under the jurisdiction of FERC (except Alaska, Hawaii, and most of Texas). Retail transactions are under state jurisdiction.

In many regions of the country, there is a market for wholesale energy such that all generators, regardless of the owner, are treated the same with respect to transmission access. These generators are used when companies compete for sales by making bids for their price for energy to an independent entity that decides which generators will run every hour of every day. The important thing to understand for this discussion is the impact that the market can have on the reliable operation of the grid. In order to ensure reliability of the grid, there must be enough generation to match the load, local reactive power must be available, and there must be some reserve generation for the things that can go wrong. And things do go wrong, almost every day, such as generators failing, errors in forecast of load being made, important transmission lines being lost, and so forth.

But does a generator that has not won the bid for the day get paid to be a reserve unit? If a generator doesn't get paid for being on call, what are the chances that the unit will be available if suddenly the grid operator asks for help? And what happens to a generator that won the bid but then failed to be available during the peak load on the next day? All these events can easily lead to problems. The point here is that the rules for the market have got to be carefully constructed to ensure reliable operation.

For the best example of how not to design a market, take a look at California in the late 1990s. In 1996, California embarked on a huge effort to design and implement a competitive electric market in the state. Their goal was to implement competition all the way to the retail level. The whole industry was watching, knowing that whatever California did would be implemented everywhere if it worked. I can remember spending hours reviewing all the information about the design of the market and trying to figure out how this would affect us in Virginia. I can still clearly remember our chief system operator telling me, "Don't worry about it. It will never work."

What they implemented in California was a very complicated package of rules along with a time frame for implementing various stages of the new competitive market. Of course, the thinking was that the cost of electricity to the final consumer would be reduced, as competition would expose "bloated utilities" and enhance innovation. Some major components of the design required the incumbent utilities to sell off 40 percent of their generation to unregulated generation companies, establishing a day-ahead market for energy run by an independent market maker called the *California Power Exchange*, which (1) allowed out-of-state energy to be priced higher than in-state-produced energy, (2) set a fixed rate for retail sales of electricity for which the incumbent utilities were required to deliver, and (3) did not allow the incumbent utilities to enter into long-term purchasing contracts for generation.

The outcome of this design is now history. Marketing firms found huge loopholes they could exploit in order to play games with generation availability and, hence, control prices. This resulted in prices at the wholesale level skyrocketing to levels that had not been seen before. For example, electric energy could be produced at an incremental cost (only looking at the fuel cost of the generator) of about 5¢ per kilowatt-hour.

This would be equivalent to $50 per megawatt-hour (1000 × 5¢), which was the minimum quantity sold on the wholesale market. What we saw, though, was a wholesale market that exceeded $250 per megawatt-hour almost every day and that was often at $1,400 per megawatt-hour or more.

Rotating load curtailments were used on several days to protect the grid from collapse. At the same time that wholesale prices were higher than ever, retail prices were fixed by the rules established by the state. This was the one major flaw in the whole plan, since there was no demand response to the high prices. Therefore the local utilities had to pay more for the energy at the wholesale level than they were allowed to sell it for at the retail level. This led to near-bankruptcies of the utilities in the state. Luckily, and a credit to the grid operators trying desperately to keep the lights on, it did not result in a grid collapse. My chief system operator was right.

Solar Magnetic Disturbances

Every so often the sun discharges huge solar flares that impact the earth's magnetic field. For unknown reasons, these disturbances seem to follow a cycle of about eleven years. In March of 1989, a disturbance caused the Hydro Quebec grid to collapse, leaving some six million people without power for up to nine hours.

The problem for the electric grid is that the material from the sun hitting the earth's magnetic field causes a voltage differential in the ground. As I've mentioned before, the voltage causes current to flow. This wouldn't be a big deal if we didn't have transmission lines spanning large areas of the earth, grounded at various points along the way. What happens is that the electric current, finding the path of least resistance, flows through the transmission grid from points of high voltage to points of low voltage on the earth's crust. This flow, which is a very low frequency, causes additional heating in devices on the grid, especially transformers, and also causes protective relay systems to operate. If allowed to continue, these unwanted flows can permanently damage equipment such as large transformers, which take a long time to replace. We really don't know what the maximum size of a solar storm could be.

Perhaps we've just been very lucky since the time the grid was developed. Some of the late-night doomsday programs love this kind of scenario.

What Can We Do to Keep the Grid Under Control?

The grid successfully performs its function of taking energy from generators and delivering it to substations/loads based on a number of very complex factors that work together. The main things that we can do to ensure grid reliability are as follows:

- establish design rules and build the grid
- establish operating rules and operate the grid
- establish rules for the market and for good performance by wholesale purchasing/selling entities
- establish rules for generators and for good performance by generators
- utilize quick-acting automatic protection systems
- ensure maintenance

> These are the main factors that will ensure grid reliability.

Establish Design Rules and Build the Grid

I will start with the design of the grid. In this section I will be presenting the following topics:

- basic modeling and designing of grid elements
- establishing rules for acceptable levels of resilience
- some examples of acceptance criteria
- how a decision on acceptance criteria can affect building of infrastructure
- assessing stability with an example of an actual event
- designing to survive a solar magnetic disturbance

As mentioned in chapter 1, there are engineers who plan the development of the grid. This is done in long-term studies, usually at least ten to fifteen years out. Their inputs include all new transmission and generation construction projects presently in the plan, load growth estimates along with estimated locations for the growth, new generation plans, any generators (or major loads) that may retire, and any consistent issues that have come up in the operation of the system. They also use similar forecast information from neighboring utilities.

What the engineers do with all this information is they build a huge model of the electric system that they can use to test multiple scenarios. The scenarios tested include a number of outages of major transmission lines and/or generators in various combinations. What they are testing to determine is if the flow of electricity and the voltages available at critical points on the grid for these conditions are within a defined range.

There are rules that the industry folks have developed over their years of doing this that define the acceptable range of results. These rules were developed under the auspices of an industry group called NERC (North American Electric Reliability Council [now it is a corporation]). NERC was formed after the northeast blackout of 1965, when the federal regulators were threatening to take a more active role in the operations of the grid. The industry formed NERC to demonstrate that they could monitor and control themselves using peer pressure. Standards were developed by experts who worked in the industry, and all players vowed to comply. The point I want to make here is that there are standards that the industry follows that describe the limits that the grid must be designed to meet. But consider this: as the criteria for design include increasingly severe situations, utilities must build more infrastructure.

I will present some examples of planning issues to help explain the kinds of decisions that are made. The easiest-to-understand example is the rule that the system must be designed in such a way that the loss of any single element (e.g., a transmission line or a single generator) does not cause the loss of grid stability, nor does it require any manual load shedding to avoid overloading other transmission elements. The golden rule for planning and operating the grid is this: the system must be able to survive a single loss of an element of the grid. This is called a *single contingency*.

When a networked transmission line is lost (tripped), the current that was on that line is instantly transferred to other, available paths. This should imply that designing a system that relies on transmission lines being loaded above 50 percent of their capability is risky, since the loss of nearby lines would result in an immediate increase of the current on the line. As a footnote to this design concept, I'll mention that in the early nineties, the opponents to building transmission lines said something like, "Their average loading today on their transmission lines is below 50 percent, which proves that the lines are already overbuilt."

I would like to provide a little additional detail on this single-contingency concept. Consider as an example a situation where studies show that a generator is needed for support in the case where a single transmission line has tripped off-line. Now the question that must be answered is if there is concern about the loss of the line when the generator is not available. Generator availability in the 90 percent range would be considered pretty reliable, but that would leave 10 percent of the hours (876 hours per year) without the local support of the generator. Would the rules allow the grid to be at risk for that many hours every year? Will the rules allow for operator intervention in this case?

The simultaneous loss of a transmission line and a generator can be a very severe contingency for some systems, especially when the reactive component of electricity is considered. Operator intervention, which would include something like redispatching generation, or shutting off large segments of load, may be necessary to survive this under peak loading conditions. The decision of which way to go on this point can be a daunting one to make, as the decision to design the grid so that this contingency does not require load reduction may result in the need for a new transmission line. The outcome may be that the simultaneous loss of generator and line requires load shedding, which could go on for days, weeks, or even months if the line is severely damaged and the generator is not available. And to make matters worse, the load shedding could be required over a large area of load, such as a large city.

Of course, there are many potentially worse situations than the simultaneous loss of a generator and a transmission line (for example, the simultaneous loss of all the transmission lines on a right-of-way) where the rules would say that a load shed is acceptable, but the grid must be able to survive. Decisions made that determine the extent of operator intervention assumed in these cases become critical to the amount of infrastructure needed. That is, if the assumption is made that the grid must survive without any operator intervention, then there will be a need for additional lines or other elements.

Planners also look closely at the possibility of an unstable generator for various conditions on the grid. Previously I mentioned what happens when an electrical perturbation of some kind (for example, a short-circuit on a nearby transmission line) occurs and the generator "wobbles" a bit with the grid. This happens regularly, and the planners must ensure for the operators that the wobble is damped. If the wobble isn't damped, the generator will normally be tripped off-line, hopefully before there is some damage to it.

The electric company for which I worked had an event in the middle of the night sometime in the early 1990s that I still shudder to think about. A large transmission line suddenly tripped off for no apparent reason, and this created instability between our pumped storage units, which were pumping water at the time, and other units on our system. The wobble I was talking about earlier didn't dampen; it got bigger and bigger! Fortunately, our shift supervisor on duty immediately realized what was going on and ordered the pumping units to be tripped off-line one at a time. There were five units pumping at about 400 Mw each, to put some perspective on the size of this event. After the fourth pump was tripped, the system calmed down.

This perturbation was felt throughout the eastern grid. Had our planners missed something? I don't think so. You simply can't study every possible configuration of the grid. The good news is that after this scare, automatic controls were put into place on the pumps that prevented this instability.

I mentioned solar magnetic disturbances (SMDs) earlier. There are some basic things that the planners can do to help the grid survive these events. One thing that has helped is that resisters were installed into the path along which the SMD-caused ground current would flow. This is done where the transformers are grounded. By doing this, the current flow from the earth's crust through the grid is reduced. Another thing that should be done is to make sure the protective relays are designed to take into account the possibility of SMDs. Does doing these things prevent the next SMD from causing a grid collapse? Since we have no idea how intense one of these events can be, there is really no way to know the answer.

I think the grid can survive with little trouble the highest-magnitude SMD that has been seen since the grid was developed. Also, if an extreme event occurred and managed to cause a collapse, the real questions relate to how much damage has been done and how quickly grid operators can get the grid back up. My guess is that the damage would be minor and localized, and that the grid could be completely back in a few days at worst. There is no reason to lose sleep over an SMD-caused grid collapse.

Establish Operating Rules and Operate the Grid

The operating rules of the grid were established by the industry group mentioned above, NERC, after the 1965 blackout. Operating rules cover a wide variety of issues, mostly focused on a time horizon of present time ("real time") to seven days from present time. I won't describe all the rules, as there are too many to go into, but I will describe those that are most important to the reliable operation of the grid. I will discuss in detail the following:

- rules for balancing generation and load
- operating to the single-contingency rule
- establishing regional entities that see the big picture
- other operational rules

Rules for Balancing Generation and Load

The control mechanism for the portable generator in chapter 1 was a governor that maintained a constant speed. This mechanism may work for small systems, but it does not work for the grid. No single generator can control the frequency since the system is so large (the peak load of the eastern grid in 2007 was 638,500 megawatts). So the control mechanism for matching generation to load is not one that relies on frequency control at all. Matching generation to load is done in a distributed control scheme, wherein defined areas (subsets of the overall grid) must match their generation to their own load. The idea is that if each area is in balance, then the whole grid will be in balance. Or, said another way, if some areas are generating less than their load, then others will be generating more than their load—so the grand total of the entire grid will be pretty close to balanced and the frequency will be close to 60 Hz.

The grid is divided into these areas by definition. That is, it is not actually "divided," but by agreement with neighbors entities that control the grid define the equipment that will be contained in each area. Prior to 2005, these areas were called *control areas*, but now they are called *balancing areas*, which is a bit more descriptive. The rules state that all grid equipment, load, and generators must be within a defined balancing area that is controlled by a balancing authority (BA), so balancing the entire grid is controlled by a number of BAs.

These balancing areas typically fall where ownership of the grid falls, although there are many exceptions. They are also closely aligned with geographic areas, such as the certificated areas that the states have defined for IOUs, although again there are numerous exceptions. At my last count, there were 105 BAs in the eastern grid. Within a BA there can be multiple generation companies, transmission owners, and distribution companies. (However, Texas has only one BA in its grid.)

Each of these BAs is connected to other BAs by transmission lines that cross the BA boundary and then connect to other BAs. In fact, most of them have multiple connecting lines since the grid is so highly interconnected. Electric current freely flows on these lines, as the law of physics dictates, so think about a large geographic area with lines connecting to other areas, such as in diagram 2-1. To simplify the example, I show only one line connecting BA-1 to each of the three other BAs. Each of these lines has metering that can give a real-time signal of the current and voltage, and therefore the power flow is a known quantity. (As I go through this example, keep in mind that there is nothing other than the law of physics controlling the flow in each line. Also keep in mind that the example is only showing power flows between BAs. Internal to each BA are power flows of generation to load that are probably much higher than the flows on the tie-lines.)

In my example, line 1 has 100 megawatts (Mw) flowing out of BA-1; line 2 has 50 Mw flowing out of BA-1; and line 3 has 150 Mw flowing into BA-1. So if we simply total all the outgoing flows with all the incoming flows, we can determine if the BA is balanced. In this case it is in balance, since all the flows in and flows out total to zero. What is interesting here is that the system frequency doesn't matter; also, the total load/generation in the BA doesn't matter. All that matters is the sum of the flows, which in this case is zero, showing that there is no balance error.

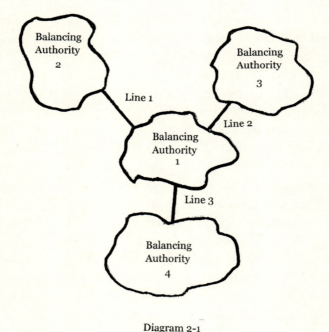

Diagram 2-1

Balancing Generation and Load

I will take this example one step further and introduce what happens if the generation and load are not balanced in the area. Assume the flows in diagram 2-1 are 100 Mw out on line 1, 50 Mw out on line 2, and 160 Mw in on line 3. There is a balance error of 10 Mw. Since the power is flowing into BA-1, the generation in the area is short by the 10 Mw error. To remedy the error, the BA-1 controller sends a signal to certain generators equipped with control capability (called *regulation control*), signaling to those generators to increase their output slightly in order to make up the 10 Mw difference. Those generators increase output slightly, and then the balancing error disappears.

This process is continued day and night at all BAs. These controls are done automatically, with a balancing error calculated every few seconds and with signals going out to generators as needed. Keep in mind for this example that the frequency of the grid and whatever else is happening in other balancing areas doesn't affect what this particular balancing area is doing. Even the total load in BA-1 is not a factor. (A little later I'll bring up an important exception to that concept.)

I would like to go a little deeper in order to explain how wholesale purchases and sales of electricity are implemented on the grid. Starting with the example above, also include the datum that one of the utilities within BA-1 in diagram 2-1 has purchased energy from a utility in BA-4. Just to make the numbers work out, the utility in BA-1 has purchased 10 Mw of power for one hour (10 Mwh of energy). How is that transaction going to actually happen on the grid?

The process to implement a wholesale transaction is really quite simple. What happens is that the balancing authority in area 1 inserts a "schedule" of 10 Mw received for the hour into their computer calculation, and BA-4 inserts a schedule to deliver 10 Mw for the hour. This means that BA-1 expects to receive the 10 Mw, so the flows on the lines in the example above are as follows: 100 Mw out, 50 Mw out, and 160 Mw in, indicating that BA-1 is now in balance. That is, BA-1 expects to receive 10 Mw from the grid, and they do receive it, since the sum of all the flows equals a 10 Mw receipt. (Keep in mind that the 10 Mw schedule from BA-4 to BA-1 would not necessarily flow on line 3 as in this oversimplified example. It would more likely flow on all the tie-lines into BA-1.) Transactions like this are constantly being implemented by almost all the BAs on the grid.

I need to introduce an important control factor called *frequency bias*, which helps to maintain the reliability of the grid. Remember that all the BAs are doing what they can to maintain generation and load balance in their areas, although there is no overall controller. System frequency is fluctuating around 60 Hz, although no one is trying to control the frequency. When a large generator trips (which, as I've said before, happens almost daily on the eastern grid), the frequency on the entire grid decreases. All the BAs in the grid (with the exception of the BA that lost the generator) will experience an increased outflow because their loads will decrease with the lower frequency—so they will all be overgenerating. With the controls I have described so far (that is, controlling generation to reach a net zero on a BA's tie-lines), there is a concern that the frequency could continue to decay, since this outflow from each BA would be curtailed. So there is a need to add a factor that allows for this flow.

In addition to the increased outflow due to the reduction in load, there is an increase in generator output due to the lower frequency. In chapter 1, I mentioned that the governor in a portable generator could be set to maintain a target speed. It turns out that most of the grid's large generators have that capability. So as these generators sense the frequency decline, their governors can (automatically) boost the output of the machine, thereby helping to arrest the frequency decline. This governor boost reaction to a low frequency happens with most generators on the grid, and it happens within the first minute after the loss of generation anywhere within the grid.

Backing down generation throughout the grid is not desirable when frequency is low, as generator boost during low grid frequency is a good thing, so therefore the frequency bias is included in the programming for all BAs. Frequency bias allows for some additional power output on a BA's tie-lines proportional to the grid frequency. The frequency-bias factor doesn't provide for the grid frequency to recover to the preferred frequency, but it does provide assurance that the decay in frequency will not become worse. To put into perspective what this factor amounts to in power output, I will take a look at the standard. The standard requires that the bias setting should be no less than 1 percent of the system load for a change of 0.1 Hz. For a system of 20,000 Mw, the bias would be set at no less than 200 Mw, which, when spread over a large number of generators, is not that much extra generation. On the eastern grid, a change of 0.1 Hz is huge; this much frequency change is very unusual.

What about the BA that just lost the generator? The standards require the deficient BA to make up the loss in generation, in any way possible, within fifteen minutes of the unit trip. There are several options for doing this, and some of them are better than others. The fifteen-minute requirement pretty much rules out starting any new generation, unless the BA has some small gas turbines available that can take significant amounts of load in this short time frame. The larger gas turbines that have been built in the last ten years are a lot more efficient than the older, smaller turbines from the 1960s and 1970s, but they take too long to get up to speed and also take load to help out in this situation. Any generator within the BA that is online and available to produce more power can be used, but often there is not enough unused capacity available in the units online to replace a large unit. Hydro units are excellent for this purpose if they are available and not already running, since they can get up to full load in a few minutes after being started. Also, the BA can have an arrangement to buy or even borrow the needed energy for some period of time from another area until the BA can start another large generator. Typically, this kind of arrangement (called reserve sharing) is done ahead of time so that it can be quickly implemented at the time it is needed. But the important point here is that somewhere in the grid, generation must be increased to make up for the loss of generation that is the result of the unit trip.

This discussion about balance error and BAs is very important to understanding how the grid is controlled. To summarize: The balance error is calculated based only on a net sum of all the flows on all the tie-lines for a single BA, adjusting for all wholesale transactions across the BA border, which are entered into the computer as schedules and are then adjusted slightly again based on grid frequency. This calculation yields a number that may be positive or negative. The control of generators within each BA is performed based on that calculated factor. We call this balance error value the *area control*

> The balancing areas control the frequency on the grid by each calculating and controlling to its own area control error (ACE).

error, or ACE. If it is positive, then it indicates that the BA is generating more than necessary. In such a case, some units will be directed (automatically, of course) to reduce their output. And if it is negative, then the BA is generating less than necessary and some units will be directed to increase output. The standard established by NERC gives guidance for measuring how well each BA matches its generation to its load. The algorithm is complicated, but the point is that some balancing error in real time is expected and allowed. It would be virtually impossible to maintain an exact balance all the time.

If anyone looked at a grid frequency trend recorder for the last couple of hours on most days, that person would see a wavy line that goes over 60 Hz and under 60 Hz continually. When the grid is at 60 Hz, we say that the generation and load are balanced, which really means that for an instant of time the load is met exactly by the generation and the frequency happens to be 60 Hz. If the load and generation are said to be imbalanced, what we really mean is that the frequency has deviated from 60 Hz. If the frequency is higher than 60 Hz, then the generation output at that instant is too high. Conversely, if the frequency is lower than 60 Hz, the generation output is too low. Fortunately for us all, electrical equipment (generators and load) is designed to handle some deviation in frequency. More important, the inherent characteristic of most of our load devices is that the energy required at a lower frequency is slightly less than the energy required at a frequency of 60 Hz. The reason I say this is fortunate is that the small imbalance between generation and load is simply absorbed by the ability of the load to increase or decrease with the frequency. Remember that all the electricity generated is used instantly. This allows the system to be in equilibrium and yet stable, which it is, generally.

It is interesting to note that this system of multiple BAs controlling to their own balancing equation makes it very difficult to identify in real time which area is causing a problem. Say for example that the frequency on the grid has been low all day. Each BA in the grid has been supplying the frequency-bias energy as required, and by now they all want to know who is getting this energy and when they are going to get the low frequency straightened out. We've seen this happen many times. There are at least two obvious ways that this can come about, and either one is tough to identify.

First, say there is a BA with a metering error on a tie-line with a neighboring BA. Actual flow in this example is 100 Mw, whereas the meter shows 10 Mw. The BA uses that meter, along with all the other tie-line meters, to calculate a total flow in or out of the area, so whatever it is that they expect to see is off by 90 Mw. In effect, the BA is receiving 90 Mw more than the computer realizes. This goes on until someone catches the error. This meter error would result in the grid frequency being a bit slow, since there is an undergeneration in the grid of 90 Mw.

Second is a case where one BA thinks there is a transaction from a generator located in another BA of 100 Mw, but the other BA thinks it is 10 Mw. The outcome is the same as in the first example. This imbalance could simply be the result of an entry error somewhere, or, as we saw in the late 1990s, it could be the result of a dispute wherein the two parties simply disagreed and left it at that. I know it sounds ridiculous, but this actually happened for several months, as I recall. During that time, every other BA in the eastern grid was supplying the makeup energy to boost frequency and was not being paid for the energy. The energy supplied over several months turned out to be thousands of megawatt-hours. What several BAs finally did was simply to remove their frequency bias until the matter was resolved. By removing their frequency bias, they were no longer supplying energy to the grid for which they were not being paid, but at the same time they were putting the grid at some additional risk.

I have tried to provide a basic description of how generation and load are balanced on the grid in real time. This discussion can be expanded to include the processes that take place to plan the present day, the next day, and the upcoming week. Without going into detail, I'll say that each BA would be responsible for developing viable operating plans for the time horizon of one week in the future. This plan would include load forecasts, generator availability plans, outages, purchases, sales, and any other inputs that may need to be on a BA's radar.

The rest of the discussion on how the grid is controlled will focus on methods of preparing for and dealing with unplanned events, and dealing with the competitive market.

Operating to the Single-Contingency Rule

As I mentioned before, our standards require that the grid should always be capable of surviving a single contingency. Obviously we don't want a blackout every time a transmission line fails! We lived with this rule for years without the capability in real time to actually be sure we adhered to it. We depended on studies that were done by the design people with models that closely represented the system as it was planned. These studies examined how the grid would respond to a large number of events, which may include lines out in addition to generators out. We call these "off-line" studies because they don't represent the system as it may exist at any given point in time. The hope was that the different cases "bounded" the issue for the operators. By "bounded," I mean that whatever situation exists in real time is not as severe as that in a study that was run off-line. Was there ever a time when the system was one contingency away from a blackout? In most cases the answer is certainly no, but there were days about which we may never know for sure.

What needs to happen to be sure this rule is met is that a calculation must be performed to verify that nothing gets overloaded after any single contingency. Ideally this calculation would be done from a starting point of existing conditions (flows, generators online, etc.) on the grid. For example, the model should take out one transmission line from the existing conditions and then recalculate how all the flows redistribute through all the possible paths from generators to loads. Then we would need to verify that all equipment can withstand the new flows that instantly appear after the modeled contingency occurs. There are hundreds of transmission lines for a typical owner as well as probably a hundred or more generators in each owner's areas. And by the way, this calculation needs to be performed at least hourly to take into account changes in generation and load. Also, as this is being done, we need to examine what the voltage will be at any substation on the system. So of course we need a computer to calculate this for us in real time.

As computer technology improved along with the accuracy of the grid models, it became practical in the early 1990s to do some of this analysis online. By "online," I mean using actual system conditions for the model. In other words, a computer models the electric grid and has all the real-time data such as generation outputs and load—the lines that are in and those that are out. Once all those data are available, the computer then performs a calculation for each single contingency and compares the results against known limits of the equipment. In fact, the larger transmission operators in the United States today probably run all single-contingency tests every few minutes. When I retired from Dominion, we were running over seven hundred contingencies every five minutes.

Once we have the results of the single-contingency analyses, then we are ready to say that the system is safe and we are in compliance with the rule. But what do we do if a single contingency is shown to cause an overload somewhere? There are a few things that can be done in this situation, which, by the way, happens very often. The first choice and the most common solution is to do what is called a *redispatch of generation*. Another choice that may be used on occasion is called a reconfiguration of the grid, where changes to the network interconnection are made. In chapter 3, I will go into more detail and provide examples of how each of these options work. The point here is that for a reliable operation, an operator needs to know where the problems exist in the grid and must respond appropriately.

So far I have been examining the analysis of the system as is done for multiple single contingencies. How about the capability to examine the system while taking into account some maintenance work planned for the next day, or taking into account a plan for a large generator to be off-line and seeing if any new problems arise? These studies are called what-if scenarios by the operators. Most transmission-owning companies do have the capability to perform these studies. In fact, most of the transmission companies have a process in place that would look at system conditions along with the expected load for a week ahead of time and determine if it would be okay to take a line out of service for maintenance. If any limits are exceeded, the program will warn the operator that there is a problem with the planned outage. This would be an alert that an outage may not be acceptable.

Several folks for whom I gave tours asked the great question of what happens if two events occur simultaneously. This would be called a *double contingency*. Actually there are several double contingencies included in the studies I discussed above. In general, double-contingency events can be especially troublesome if the two elements lost are fairly close electrically on the grid. As I mentioned before, for most tripped-line cases no load is lost immediately (although there might be some lost eventually). The power flow to the load just finds alternate paths. NERC rules require that after a single contingency occurs the operators have a thirty-minute window to prepare their part of the grid for the next single contingency. This is reasonable, since it would take some time to recover from the event, do the contingency studies on the new configuration, and respond to any exceeded limits in the new study. But there is an amount of time after an event when the system is at some potential risk.

NERC standards do not require operators to be prepared for all double contingencies. Why not? Consider that there are one thousand elements in the area of concern. The transmission operator has reduced that number down to, say, five hundred single contingencies that could be cause for concern. In effect, the operator has decided not to study losses of small generators, lower-voltage lines that are radial, or lines that take the load with them when they trip. The question is, how many double contingencies are there for the five hundred elements? The equation for this probability (500 taken two at a time) results in 124,750 double contingencies to study. This is a number of studies that is simply not practical for our present technology to perform. The fact is that there may be hundreds or even thousands of these double contingencies that would show serious problems, so the operators would not be able to deal with the results of the analysis anyway. There are companies that are looking into ways to simplify this by not looking at two elements that are not really close to each other based on some tests that can be administered. Anyway, maybe someday there will be a way to do this. But for now, if we have an unexpected double contingency, we are going to have to act fast.

Establish Regional Entities That See the Big Picture

After two significant grid blackouts in the western interconnection in the summer of 1996, NERC embarked on a program to initiate a newly defined entity with a view of the grid that is at a higher level than the view the transmission owners have. The effort to develop the role, authority, and description of this entity was assigned to an industry task team consisting of primarily operations personnel from across the United States and Canada (I worked with this task team). It also included a few representatives from the marketing community.

After Congress issued the 1992 milestone act stating that the transmission grid must be operated as a common carrier, and once the subsequent FERC Orders 888 and 889 were issued (in 1996), all hell broke loose in operations. I am not saying that the FERC orders were a bad idea; in fact, in the long run they have improved reliability. But the impact in the early days was disastrous. In hindsight, if the industry could have spent a couple of years developing the tools needed to operate under this new model, a lot of problems could have been prevented. As I recall, though, FERC didn't want to hear about any companies needing more time—and frankly, if FERC had allowed more time for development, grid operators probably wouldn't have been able to get everyone on the same page without actually experiencing the problems that we eventually encountered.

The issue, in a nutshell, is this: Prior to the new rules, wholesale transactions were done by the system control operators in a simplified method, and it was fairly easy to keep track of them. Once the market opened up and a whole new player was in the loop (the marketer), there were far too many transactions to keep straight. At the same time, we were using the simplistic accounting methods we had always used. It became apparent early on, and especially after the western blackouts in the summer of 1996, that the operators were overwhelmed with transactions.

Add to this dilemma the fact that a transaction could be legally arranged that had large current flows through a part of the grid whose operators had no knowledge of the transaction. Why does this happen? It is the difference between the laws of physics and the laws of contracts. The problem is that those pesky electrons don't care to follow the contract path. Instead, electricity follows the path of least resistance. Without getting into physics, I'll say that the path of least resistance in a network involves many parallel paths. That is, all the electrons don't flow on one path through the network (the contract path), but instead they divide up and take numerous paths.

See diagram 2-2, below, for how this might work. I have included this diagram to help illustrate this very important concept of networked systems (like our grid). In this example, a generator in BA-S (the source BA) has purchased contractual rights to transmission through the transmission owners in BA-S to get to the border of BA-S, and has a purchaser within BA-3 who has transmission rights to import power. (The contract path is actually through transmission owners, but I've simplified the example to show the BAs, which in most cases would be the same entity as the transmission owner.) In this example, the transaction is for 100 Mw for a period of time. We know that 100 Mw leaves BA-S and that 100 Mw arrives at BA-3, but the flows in a real-life case are even more complicated than the example I've provided in diagram 2-2.

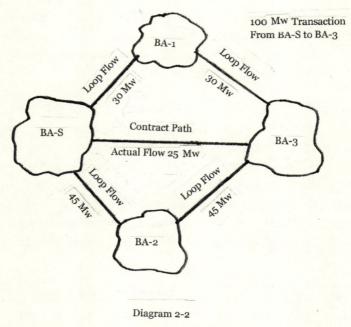

Diagram 2-2

Contract Path vs. Actual Flows

The actual path invariably would involve some transmission owner that had nothing to do with the contract for delivery but was carrying a significant portion of the current. In this example, look at the transmission owner in BA-2. The operators for this transmission owner have no control over this contract and don't necessarily even know it is taking place, but the largest portion of the flow is through their part of the grid! This is merely a consequence of a network system, but it is extremely tough to deal with when lines are approaching overload and the operators don't know where the flows are coming from. The combination of these flows, called loop flows, would be the result of multiple transactions occurring virtually anywhere in the grid. And since the transmission operators who are experiencing problems have nothing to do with the contracts in the first place, they also have no recourse to stop the transactions from flowing through their lines.

To put some additional perspective on this problem, I'll mention that the overload on a transmission line could be the result of ten, twenty, thirty, or more different transactions. We call the problem point in the grid a *flowgate*, which is a term used by those in the industry to identify typical points in the grid that are constrained often by these loop flows. Making a detailed analysis of the many transactions and the makeup of the grid at the time would be necessary to approximate the source of the flows seen on a flowgate. But since this was not done, and since the operators were left with no means of controlling these flows, huge problems with controlling the grid were encountered, which led to the western grid collapse in the summer of 1996.

Why didn't this problem show up prior to 1996? This is a great question. The answer is really one of scale. Prior to 1996, we did transactions all the time, and there were always loop flows (in fact, to be precise, a generator near a border of one utility would impose loop flow on its neighbor without any transactions). Occasionally there were problems caused by these loop flows, but it was usually within a small set of companies that knew what was going on and could work out solutions. After the new rules came about that opened up the grid, the number of transactions increased exponentially, and these transactions went longer distances as well. It was this factor of a large number of deals over longer distances that made it very difficult to unravel in real time when the grid was at risk.

What was needed was a mechanism that tracked all the transactions going on in real time and that also calculated the impact of each transaction on each flowgate. And maybe even more important was having an entity with authority to do something about the problem when the combined total of all the transactions threatened grid collapse. So we created the role of a reliability coordinator (RC) that had the authority to order the curtailment of transactions, even though the RC was not a party to the contract. Of course this concept had to be filed and approved by FERC, but FERC was very supportive, as it was clear to all that this role was needed. After implementation, the RC would have the big-picture view that the individual transmission operators or balancing authorities simply did not have. The development of the systems that logged and tracked transactions and calculated their impacts on flowgates was done at NERC—and it took some time to get these systems straight. The industry initially developed about twenty RCs across the United States and Canada in 1997, although that number has since been reduced to about sixteen.

Other Operational Rules

There are many other rules, expressed in industry standards, that are important to the reliability of the grid. I will just briefly mention a few examples to show what it takes to keep the grid in operation. For instance, there is a standard stipulating that all operators have to be certified by NERC, which means that they have to pass an exam to become an operator. Also, they must speak English in all conversations outside their own company. These operators also must have thirty-two hours per year of simulation training on blackout scenarios. There are also standards that require the sharing of information, such as planned outages, so that it is possible for neighbors to analyze the grid using good information about equipment outside their own system.

Establish Rules for the Market and for Good Performance by Wholesale Purchasing/Selling Entities

I showed earlier how not to set up a market and how a bad idea in the market design can cause a huge risk to the reliability of the grid. The main point is to have rules in place that ensure that the reliability of the grid always comes first. Although there are those who blame competitive markets for most grid problems, with the right rules in place the grid should be fine. In this section I will discuss the following:

- the fact that wholesale competition was a challenge to reliable operations
- examples of problems with implementing the new processes
- how competition enhances reliable operations now
- the fact that investments in generation rely on rules that enhance long-term planning
- the fact that transmission design and infrastructure development are ensured by rules

To those who say that competition was the harbinger of several blackouts, I will say this: in the early days, after FERC ordered the implementation of open access, the industry was simply not ready for open access. FERC was probably tired of waiting for the industry to get ready. Frankly, I think that they didn't believe the big utilities when they warned of problems to come. We went through several frustrating years of working in an environment where Enron and other marketers had the ear of the FERC and we old utility guys were seen to be just crying wolf because we were afraid of competition. Maybe so, but we were also afraid that reliability was going to be impossible to maintain with all these new players and new rules. But we were the dinosaurs, and the model that Enron and others propounded was the future. I remember Enron's mantra: "You don't need iron." This meant that there was no need to own generation; one just needed to be smarter than all the other players and to trade the output of generation. The utility guys were DOUGs: dumb old utility guys. Enron's idea of "smart" was finding the loopholes in the rules and playing the game for the most money. This didn't turn out well for them in the long run.

After the western blackouts in the summer of 1996, those of us in the operating arena realized that we had to get our act together. I've already discussed what was done with the new RC role, and the calculations that were needed in real time. All of this had to take place with minimal impact on the market. There was a time when all the new tools and new types of transactions brought havoc to operations on the eastern grid. What happened was that on a particularly hot day, the price of energy on the grid went through the roof, as it did a lot in the late 1990s. The marketing community was extremely active with trying to make money. There were more transactions than a certain large utility in the eastern grid could keep up with—probably hundreds of transactions an hour. The tools were only partially automated, although the rules were clear. In fact, the rules stated that neighboring BAs had to verbally verify the scheduled flow between them anytime there was a change.

The transmission owner that I mention in this story reached the point of being so overwhelmed that its operators quit answering the phone. At that point, the question for those of us who were neighbors was, "What do we do now?" We also had large numbers of transactions to implement, but by rule we couldn't change the scheduled flow without a verbal verification. So we had a conference call (without the large utility, of course, since they wouldn't answer the phone) and quickly decided that the only thing to do was to continue the scheduled flows we had until we could verbally agree to change them. This actually went on for several hours. Needless to say, this all took a long time to sort out in accounting, and there were some unhappy marketers who missed a great opportunity.

A lot of improvements have been incorporated since the early days, and the process of improving the mechanisms and rules continues to this day. I believe that the system we have developed has evolved to the point where wholesale competition enhances reliability. The reason I say this is that the ability to seek help throughout the entire grid during times of great need is a big step forward. The eastern interconnection is so large that there are different weather patterns and even different time zones, so there is always some diversity in demand. If the East Coast is experiencing an extreme cold spell, there are companies to the southwest that probably have their reserves. The marketing system that evolved will quickly identify what is available, and other systems will verify the ability of the grid to deliver. This simply was not the case in the early 1990s.

Another concern related to the coming of competition in the electricity business was investment in generation. What was quickly realized by many in the industry was that we were in for a roller-coaster ride for generation investment. When energy prices were high, there would be plenty of investment, but when prices were low, nobody was going to invest. With a fairly steady load growth, around 2.5 percent annually in the Southeast, there was going to be a continuous need for new generation. Look at it this way: a 2.5 percent load growth on a 20,000 Mw system is 500 Mw every year. With this kind of load growth, it is dangerous to let too much time go by without building some new generators. What seemed to be happening in the latter part of the 1990s was that state regulators were closely following California's competitive market development (before it failed) and were making it clear that they too were going to deregulate generation. This would mean that the big multiyear base-load-generation projects, such as nuclear power, just weren't going to happen. An investment like a nuclear unit needs to be in the rate base if it is going to be built. Without that assurance, a nuclear unit is far too economically risky to build. Not constructing new generation (or depending on the market to provide it) was a dangerous path to travel from a reliability perspective.

So what is a good set of rules for the market to ensure reliability of the grid? At a very high level, there needs to be an incentive for generators to be available, whether they are called to run or not. Availability of the generator is what is called *generation capacity*, and this is measured in megawatts. Early competitive systems were designed around a market where generators only earned revenue for energy (megawatt-hours) delivered. Sounds good, but it doesn't lead to a very secure grid. Why not? If generators are only paid for energy output in a competitive market, those that are too expensive to run will simply shut down when the market is low. In such a case, one might as well lay off the crews for the season and then have them show up in time for the summer. There were units that actually did this, although it was not widespread practice. The problem with this is that in order to have a reliable system, there must be some available generation reserves. And units that are supplying reserves have got to be paid for their availability, not just for energy delivered.

In addition to the reserve issue, though, there is the need for reactive support near the major load centers. There are a number of generators that are in a location on the grid where their output is essential to the reliability of the grid because they supply this reactive power. Remember the discussion of reactive power, above, and the fact that it can't be transmitted very far? The problem here is that the plants that supply the most critical reactive power are typically not the latest in efficiency, and although they are crucial to the grid, in a totally competitive generation market they will not survive. Licensing a new plant to take the place of these plants near the cities is nearly impossible.

The markets that have evolved have solved these problems to a large extent. There are location-based payments that allow crucial units to stay online while, at the same time, controlling the prices so that these units do not receive a windfall. As far as generation capacity goes, there are mechanisms that pay generators for their capability, even on days when they are not called to actually produce energy. In the discussion about the California market, I mentioned that the marketers withheld generation to keep prices up, but now the rules are more tightly written so that gaming is not as easy (for example, any unit that declares it is unavailable may be subjected to an inspection to verify that fact. It may also be subject to a deduction in capacity payments as a result of its unavailability).

Transmission investments in the mid to late 1990s dried up to the point that it seemed as if nothing was going to get built anymore. Some transmission owners even took money out of the transmission-maintenance programs. I suspect that the money was used in many cases to fund generation in the unregulated business. When the energy market was high, returns on generation far exceeded anything possible by FERC-allowed returns on transmission.

With regard to transmission-system investment, the NERC rules are very strict, stipulating that design planning and infrastructure development must be performed regularly. In addition to this, though, is the return on investment allowed in a transmission owner's approved rate, which seems to be providing an added incentive to build needed additions.

Establish Rules for Generators and for Good Performance by Generators

I want to discuss briefly the rules for grid reliability that apply to generators. These rules, established by NERC, should not be confused with market rules, which are specific to whatever market in which the generator does business. There are not very many reliability rules that apply to generators, but the ones that do apply are important. In this section I will discuss the following:

- communications required of generation operators
- automatic controls for reactive support

For the most part, the rules require the generator people to communicate with the transmission people. This communication would be about things such as facility ratings, changes in ratings, planned and unplanned outages, and specific unit modeling data that are required to operate the grid. That makes sense and seems almost trivial to have to state, but these data were not easily obtainable prior to the rules being put in place. Several of these rules establish the relationship between the generator and its transmission operator such that the generator must comply without question to data requests, and in some cases to orders from the transmission operator.

Most generators have automatic controls for their reactive power output. It is very important to grid reliability that the system remains in automatic mode, since things happen on the grid too fast for a plant operator to respond. The rules say that if the automatic mode fails for some reason, the generator operator must notify the transmission operator. The analyses that I mentioned earlier depend on good information, and these analyses assume that the generators' delivery of reactive power will automatically respond to events. If they don't, the possibility exists that the grid will not survive a single contingency.

William L. Thompson

Utilize Quick-Acting Automatic Protection Systems

Earlier, I discussed the problems that the grid faces when elements are faulted or short-circuited. I've also discussed the stability of generators when these kinds of events occur. I won't delve into the statistics of these unplanned events, but suffice it to say that they happen all too often. In fact, during a single thunderstorm there could be several events within one transmission owner's area. Over the entire grid, there are a lot of unplanned events. But an unplanned event does not have to mean that the grid is unprepared for it. In this section I will present the following:

- some examples of what can go wrong with automatic protection systems
- a comparison of grid protection with protection in the home
- examples of what the automatic protection systems do
- examples of how the protection system prevents further damage

These protective relay systems are very complex, and the people who work on them have got to be extremely careful in everything they do. The blackout of 1965 was traced to an improper relay setting on a transmission line outside of Toronto. In this blackout, thirty million people were without power for up to twelve hours. In Florida, after the mandatory standards went into effect, a relay technician inside a substation near Miami made a mistake with the relay equipment. The results were that thousands of people were without power and a $25-million fine was leveraged against Florida Power and Light.

I want to digress a bit here to make sure the reader understands conceptually what I'm talking about when I mention protection systems. Let's start with the example of the fuse box or circuit breaker panel for a residence. This is a radial system, as I've described. If a fault (short-circuit) occurs on a 20-amp circuit, the fuse or circuit breaker needs to open in order to stop electricity from going to the fault. People have designed these protective devices so that they will allow current greater than 20 amps to flow for a small period of time. This allows for the fact that many devices have what is called an *inrush current* when first started; motors, especially, have this. However, if the current exceeds 20 amps for very long, the fuse or breaker will open to kill the circuit. On a networked system like the grid, we can't use these types of protective devices. When there is a fault on the grid, a large number of lines and other devices will experience currents greater than their rating. We must deenergize a faulted line, but we don't want to trip anything that is not required to trip. So these automatic protection systems are really quite sophisticated and must be designed to allow current flows considerably higher than the rating of the line, at least long enough for the actual faulted line to be tripped.

Over the years that the electric system has been operating, engineers have devised some pretty sophisticated automatic systems that quickly disconnect faulted equipment from the grid. A human could not possibly act fast enough to save the equipment, since faults can damage equipment very quickly. I will use as an example an event where strong winds knock down a tree and it falls across the conductors. Typically what happens is that the current in the line now finds the easiest path to ground through the tree, which produces a huge arc. The automatic systems on these transmission lines will sense the ground path and disconnect the line within a few cycles (for example, six cycles on a sixty-cycle-per-second system is one-tenth of a second). If the tree brushes past the line and continues to the ground, the automatic protection system will first deenergize the line, which allows the arc to extinguish, and then reenergize the line after a short time delay. When this happens, customers may experience a slight blip in the lights.

A typical unplanned event would be a lightning strike on a transmission line. Most transmission lines are designed to handle lightning strikes up to a certain level. However, occasionally a strike will be stronger than a line is able to handle. When this happens, the transmission line is automatically removed from service for a brief time, which allows for any arc to extinguish. Then, the line is reconnected to the grid. This all can happen within a matter of cycles (less than a second). Also, the lightning strike can stress the equipment enough that something will break. In this type of case, the line is removed from service until a repair can be accomplished. A number of different things can break in these events, including the conductors themselves. The automatic protection system is designed to protect the grid from further damage when things like this happen.

Ensure Maintenance

The entire electric grid must be maintained, which in many cases requires the transmission line, circuit breaker, transformer, or other equipment to be taken out of service. In this section I will present some thoughts on maintenance, as follows:

- planned maintenance outages are going on constantly.
- sometimes the work is performed without deenergizing the circuit.
- transmission conductors are designed to sag.
- lines sagging into tall trees have been a causal factor of many blackouts.
- fines for not doing tree maintenance have solved this problem.

A system the size of the one I worked with (approximately 20,000 Mw of load) always had something out of service; in fact, there were usually many lines out of service. All the generators were not available at all times either. Operators would prefer not to have all these planned outages. But of course if the equipment on the system isn't maintained properly, it can spell trouble for the grid.

Some of the work on the transmission lines is performed hot, that is, without deenergizing the line. As long as the worker is properly insulated, there is no path to ground, so the work can be done safely. This practice is called *barehanding*, and it is routinely performed, which is a benefit to the grid. I had the opportunity to barehand a 500 kV line one time; it was pretty exciting. My hat is off to the people who do this every day.

The most important maintenance issue for transmission is clearing the trees along the right-of-way. The transmission-line conductors between the towers are designed to sag in every span. It turns out that this sag is very important and must be accounted for in the design and operation of the line. As the conductors carry more current, they heat up. As the metal in the conductor heats up, it expands, and this causes the conductor to sag more between the towers. The capacity of the line is limited by the amount of sag. Grid operators are vigilant to make sure the limits are not exceeded.

But if trees growing under the line are allowed to get too tall, none of this matters on a hot afternoon when the line sags into trees. Why would the trees get too tall? The issue boils down to this question: why spend $10 million to $20 million on transmission right-of-way maintenance when that money could earn a great return in the market for generation? A funny thing about the curtailment of transmission maintenance is that the person who orders it looks great for a few years, controlling costs and keeping earnings up, but when the trees grow up into the lines, it is left to someone else to pick up the pieces.

Tall trees in the right-of-way were major causal factors in several of our blackouts in the United States, including the northeast blackout of 2003. One of the most important rules that we have in the industry today is to maintain transmission rights-of-way. FERC can fine a transmission owner for not maintaining their lines. In fact, with FERC's ability to fine $1 million for each day of noncompliance, the utilities don't question the expense for right-of-way maintenance. This particular rule has been a tough one to live with, as there are thousands of miles of lines and it takes only one tree to show noncompliance. Many fines have been levied against transmission owners for not pruning or removing tall trees in the right-of-way, but tree maintenance is nowhere near the problem it was prior to the 2003 blackout.

In this chapter I have described the major risks to the grid and how these risks are mitigated. In the next chapter I will delve into more detail about the tools that grid operators have at their disposal when the grid is challenged in real time.

CHAPTER 3

RELIABILITY TOOLS USED BY OPERATORS IN REAL TIME

"I'm sorry, Bill, there is nothing else we can do. The units are simply frozen and unavailable." This is what I heard from a generation operator at 4:00 a.m. on January 19, 1994. Operators at other units had told me the same thing. Meanwhile, another person in the control center was being told that there was no available energy anywhere from our neighbors. The temperature in Richmond, Virginia, was around zero degrees Fahrenheit, with windchill hovering around ten below. Usually at 4:00 a.m., the morning load increase hasn't even started, and yet on this day we were running every available generator just to serve the load we had. What were we going to do when the load started coming in, typically increasing at a rate of 300 Mw every ten minutes?

Introduction to the System Operator's Tools

In this chapter I want to go a little deeper into the operator controls used to maintain grid integrity. I will examine some of the tools that the system operators can use in real time to get out of a potentially bad situation. These tools are available for system operators at most grid-control centers and can really come in handy when things go wrong. In spite of all the best planning and support for the operators, things do go wrong. The problem is that as things become increasingly severe, some of the corrective actions won't help much while other actions will not be at all popular with customers and regulators. Please keep in mind, though, that the overriding goal of the system operator is to maintain the grid. An operator may not like to enter into a load-shedding event, but load shedding is preferable over a blackout.

Operational Tools to Maintain the Grid

To frame this discussion, imagine that in spite of all the controls and systems described in chapter 2, there are still problems facing a grid operator. These are the problems that are evident on the actual day of operation—in other words, close to real time. Some of these tools are used for normal small corrections, such as using redispatch or load-management programs. But other tools, such as voltage reduction, public appeals, and rotating blackouts, are used when the situation has degraded (such as in the example mentioned at the beginning of this chapter) to such a degree that an operator needs to take control to protect equipment that is overloaded or to prevent a possible blackout of the grid. Following is a list of operator tools that I will address:

- redispatching generation
- reconfiguring the grid
- using load-management programs
- reducing voltage
- making public appeals
- instituting rotating blackouts
- doing nothing

Redispatching Generation

Redispatching generation is simply reducing the output on some generators and increasing output on some others so that the total generation is the same. When this is done, the flows through a problem area can be reduced while the entire load is still served. This is the most common form of solution to the flow problems that arise. In fact, this actually goes on all the time. Think of it this way: if our transmission system was infinite in capability, we would always run the least-cost generators to meet whatever load demand was out there. But the grid is limited in delivery capability, so operators must put limits on the generation mix that is online. For example, there is no way that the entire load in Virginia could be served by using only mine-mouth coal units in West Virginia and/or Kentucky. Not only would the lines not be able to carry the needed amount of current from West Virginia, but also, as mentioned earlier, the reactive power needed would not be available.

So the operators need to continually come up with a plan for a generation mix that meets the load demand for each hour of the day and is the lowest possible cost. But in addition to that, the plan has to meet all reliability requirements. By reliability requirements, I mean that the plan must have reserve generation, not overload any lines in the grid, account for single contingencies as discussed in chapter 2, and provide reactive power near load. It is this list of constraints (and many others) that keeps us from always running only the lowest-cost generation. For example, in order to meet the constraint of providing reactive power near load, we may have to run some older, more expensive units that are near big cities.

It is when this plan runs into problems that the term *redispatch* comes up. This is something that happens a lot, so I'm going to offer an example. Imagine that on a hot summer day the load is higher than forecast and the studies have shown that a major transmission line that is carrying power west to east would be overloaded for a single contingency. This is an unacceptable situation for the grid, so a redispatch is ordered to increase generation in the east and to decrease a similar amount in the west. The term *redispatch* is used because the new dispatch of these generators is different from the original plan. The redispatch reduces the west-to-east flow on the line so that the analysis no longer shows a single-contingency violation. This action typically results in increased cost since the generation increased in the east is running at a higher cost than the generation reduced in the west.

Reconfiguring the Grid

Reconfiguring the grid is simply making a change to the networked interconnections in such a way that the problem is relieved. It can be a simple solution that is completely within the control of the transmission operator. Reconfiguring the grid is not available very often, unfortunately, since many changes to the network would solve one problem but lead to another. But it is an important option, so I'll give some examples.

Sometimes the answer to a problem is simply to remove a transmission line from service. For example, let's say there is a small 115 kV line parallel to a big 500 kV line. Assume that both of these lines are carrying load into a large city. For the loss of the 500 kV line, the flow on the 115 kV line may become more than that line can handle. However, there may be other paths available that could carry the load, so simply removing the smaller line from service may resolve the single-contingency issue. Of course, simply removing one end of the line, making it radial rather than networked, solves this problem as well, but in this case the load that the 115 kV line serves will not have the benefit of a network source.

I've also encountered situations where asking the maintenance people to get off a line so we operators could put it back in service was the only way to resolve a problem. In these cases, we would typically have to make the maintenance people aware ahead of time that there was a risk in taking the line out of service and tell them that if some contingency occurs, we will need to put it back in service quickly. Sometimes they can do this, depending on the work needed, but most of the time this option is not practical.

In this category of reconfiguring the grid is also the possibility of energizing a capacitor when the voltage is projected to be too low after a contingency. Transmission companies typically have a number of these capacitors strategically placed around the grid to boost voltage when the loads are high. A capacitor provides the reactive power that I have discussed previously, and this reactive power gives the voltage a boost. The voltage can also get too high overnight, especially when the load is low and there are few units online to do voltage control. It turns out that large transmission lines at low load can act as capacitors and boost grid voltage too high. Many times when this happens, the operators can remove a line from service without impacting any load. Doing this reduces the voltage, taking it back to the desired range.

William L. Thompson

Using Load-Management Programs

There are a number of load-management programs being used successfully in the industry. In this classification are any contractual agreements that allow the utility to curtail demand whenever the utility asks for it. I'll mention a few examples, but I want to be sure to differentiate these from the load-reduction mechanisms that follow, such as public appeals. Customers enter into load-management programs voluntarily. Examples of these programs include programs to control water heaters and air-conditioners, wherein the utility is allowed to send signals out to devices that shut off customers' equipment. These programs are very popular with residential customers. In return, customers get a credit to their bill. There are similar programs for commercial and industrial customers to curtail their demand when the utility puts out a request. I'm going to include in this category programs that utilize generators that larger (industrial, usually) customers own. This would include generators that are typically on the industrial customer's own distribution system, on the customer's side of the meter ("behind the meter").

Most recently the concept of load management has matured into a business of its own, with third-party entities contracting with customers for the right to order demand reduction. This is priced as $/kilowatt for demand reduction, not energy (that is, not $/kWh). There have been many trial programs of time-of-day metering, where a customer's cost for energy ($/kWh) could vary during the day, but the number of customers on these programs was too small to notice inside the control room. These time-of-day programs have not shown themselves to be useful to an operator who needs help in real time.

Load-management programs are great for the economic operation of the electric system. Reducing the peak demand on just a few days of the year can save the huge cost of building new generating units. This also results in better utilization of the units that are operating, since they get to produce more energy while online. Therefore the units run more efficiently, lowering fuel cost. A few years ago, if I were to ask a system operator which he or she preferred, load management or generators, he or she would always go with generators. System operators placed more trust in having the generation to call on rather than depending on a reduced demand, although I think this is changing as these programs show their reliability. One thing to keep in mind, though, is that these load-management agreements will typically be limited to a certain number of days per season or per year, or to certain times of day. Conceivably, this could be troublesome if there is a long-term heat wave and the allotment of allowed days has been used.

Another point about load-management programs is that electric utilities have become very dependent on them. Load forecasts take into account the reduced load. Of course, since the peaks are lower, generation is not being built to serve this load. The load is out there, though. Utilities need to keep in perspective that load management is another tool that must be reliable if the grid is to be kept secure. The control and operation of many load-management programs have been turned over to marketing organizations, since the programs are tightly linked to market prices. From the system operator's perspective, the only impact of the load-management program is a new load forecast for the day, which really doesn't do much good to the operator if all hell breaks loose in the afternoon of a really hot day. In other words, after all this discussion, the load-management program may not even be a tool for the operator to use in real time. But to be clear, these programs offer a great savings for customers and for utilities, and there is a lot of potential for future programs.

Reducing Voltage

A voltage reduction is purposely done on rare occasions to reduce load. What typically happens is that the transmission operators ask the distribution operators to implement a reduction in voltage in the delivery system. Customers generally don't notice the difference, which is a reduction of around 5 percent. The lower voltage reduces the load on lots of devices, such as clothes dryers and lightbulbs. The overall effect on the system is slight, maybe 2–3 percent of load, but this option can sometimes help system operators through a tough spot, like when the load is slowly climbing past the available generation and system operators may be thinking, *If we can only get through the next thirty minutes, the forecasted thunderstorm will cool things off.* Thunderstorms have come through and saved the day many times for the system operator.

There are at least three problems with this option. First, the load may reduce for thirty minutes or longer, but most of the load will soon return, even with the lower voltage. Related to this issue is that when the voltage reduction is ended there is another increase in load as all these devices get back to full voltage. In fact, we have seen the load climb to its daily peak after returning to normal voltage (not what we want to accomplish!). The second problem is the generally negative reaction of the state regulators to a voltage reduction. Regulators look at a voltage reduction as a sign of trouble and want to come in and investigate what is going on. Voltage reduction has a negative connotation because it shows that the utility is having trouble meeting the demand. Remember that the regulatory "deal" is that the utility serves all demand in its territory in exchange for the right to earn a stated rate of return. The third problem is the possibility that there may be customers out there who are sensitive to a lower voltage. This might be a computer chip maker, a large industrial customer, or a computer server "farm."

Making Public Appeals

Anyone who has watched TV or listened to a radio on an especially cold or hot day and heard the announcer say the local power company is asking all customers to avoid using major appliances and turn off all unnecessary lights has heard a public appeal. Public appeals are a consideration when the load is expected to exceed available generation. Asking the public to reduce energy use is an effective way to get a load reduction, if done properly. Such a thing is done by notifying radio and television stations to put the notice out to everyone in the territory that the local utility is having a difficult day and needs people to hold off on drying their clothes or washing their dishes and so forth. There are really two kinds of public appeals. One type is the soft public appeal that is often used during peak days; its message is more one of "use energy wisely." This is a reminder to turn off lights and appliances if they are not needed. Another type is a more hard-nosed appeal, where energy providers admit that there will be trouble with the grid if the public doesn't reduce demand. This type of public appeal takes several hours to implement before results are seen, so it is used when the operators see a shortage coming.

If a voltage reduction has negative connotations, then a public appeal is a lot worse. As I said, the voltage reduction really isn't noticed by most customers. A public appeal is noticed by customers, and especially by the state regulators. The problem here is that we are laying it on the line for customers: "We can't serve the entire load today, and we're asking customers to reduce load." Furthermore, if the public appeal is used too often, the customers will very quickly learn to ignore the request. On the other hand, if there is not enough generation to serve the load, it is far better to warn customers ahead of time. We used the public appeal very sparingly during my career in the control center. Below, I will talk in detail about the use of the public appeal in the winter of 1994, but other than that event there may have been one or two others in twenty years.

Instituting Rotating Blackouts

In the late 1990s, the wholesale market for energy in the eastern interconnection reached astronomical levels one day. The price of energy in the eastern interconnection got up to $5,000 per Mwh (the wholesale market would typically be in the $100-$200 per Mwh range). Much like in the California situation, the retail market was still fixed at whatever rate the state regulators had decided to allow. To put some perspective on this, the rate for customers then might have been 10¢ per kWh, which equates to $100 per Mwh. Short-generation utilities were willing to pay any wholesale rate to avoid the rotating blackout, which in some cases would be the next step if a transmission company was short on generation and out of options. So these companies were buying energy on the wholesale market at $5 per kWh and selling it to their rate-paying customers for 10¢ per kWh.

According to the NERC rules, companies that are responsible to serve customer demand and can't find enough generation to meet the demand should reduce load. This is simply turning off the power to segments of customers. There are a few different terms used to describe this action, such as *rotating load curtailments*, *rotating blackouts*, and *rotating brownouts*. I always called this option a *load shed*. It is called "rotating" because the design of the load-shedding system is to rotate through as many customers as possible. Load reductions are done in the distribution system, usually by shutting off the circuit breakers at the substation, thus turning power off to all customers on that line.

Utilities learned that if the line was left deenergized for too long, maybe longer than fifteen minutes, then the load was too great on the line when the circuit breaker was closed. If the load was too high for the circuit, then the breaker had to open again and all the customers on that line would be without power until the line could be recovered in segments. After thirty minutes of no power, temperatures in many homes lower or rise to the point that the thermostat is saying to the heat pump or air conditioner, "As soon as power comes back, start up." There is some diversity in the load that we absolutely rely on in the design of the power system. Not everything that is hooked up or plugged in is on at the same time. So we try to bring back circuits we've curtailed after no more than fifteen minutes and to take off others. This way, at least some of the heat pumps won't come on as soon as power is restored.

Circuits on the distribution system are typically prioritized for this operation such that police stations and water-pumping stations would be high priority and wouldn't be shed unless the situation was dire. The lowest priority, and the loads that are shed first, are the residential-only circuits. The system that I'm familiar with was automatic. In such a system, the operator determines an amount of load that needs to be removed and the rotation occurs automatically.

The loads that are removed can be targeted to specific areas if the problem is a delivery problem such as a contingency overload, or it can be general over the entire service area for a system-wide lack of generation support. Rotations of the loads will be continued until the problem can be resolved somehow. If the problem is a severe lack of generation, then this rotation could continue for hours, until either the demand across the system drops or generation becomes available. I discussed this operation in chapter 2, under the subhead "Markets," with respect to California. The reason that the load is shed is to avoid the worst outcome: a blackout of the grid.

Implementing these rotating blackouts is an incredibly negative event in the eyes of the regulators, and this is not to mention the impact on customers. I've mentioned the compact between the utility and the regulator; a rotating blackout is a clear problem from that perspective. However, studies of the probabilities of these occurrences have shown that there is always some probability of a need to shed load. The planners who design our grid would have a once-in-ten-years probability built into their models as an acceptable level of risk. This is not to say that purposefully shedding customer load is acceptable, but that designing and building the grid for even higher levels of reliability is not justified from a cost standpoint.

During my watch in this business, we only had to resort to rotating blackouts once. However, the event is burned into my memory as if it had happened yesterday. It was on an extremely cold winter morning in January. This chapter began with one of the early-morning conversations I had on that day. We were severely short of generation in our area because several of our generators had "frozen up" overnight. No matter what we tried to do to purchase energy, we were unable to find any. Finally, with our load exceeding our generation by nearly 1000 Mw, and with the grid frequency down to 59.9 Hz, we implemented the rotating blackouts. We stopped the rotations prior to noon, as the load had decreased to a level that matched our generation. This load-shedding event in 1994 was the result of a series of missteps, unforeseen factors, and extreme weather conditions that came together at the wrong time to create this outcome. During this event we used many of the tools I discussed above, including load management, voltage reduction, and public appeals. Although they helped some, we were not able to recover a balance between generation and load.

In the afternoon, one of the local television crews wanted a live interview with someone to explain what had happened and what was expected for the rest of the day. Fortunately my boss agreed to do the interview, so my job was to deal with the logistics of the location and ensure he had the latest information. As this was going on, I ran into the *Richmond Times-Dispatch* reporter who was also in the building, looking for a story. I had talked with the guy numerous times and felt pretty comfortable around him, so I said one of the more stupid things I've said in my life: "I feel like the most stressed-out man in Virginia today." And, of course, this was the lead story, along with my brilliant quotation, in the paper the next day. To this day, I have not heard the end of that from my friends at Dominion.

Doing Nothing

Doing nothing is always an option for any problems that we face, not just in the electric business. And, frankly, this tactic has been and still is used a lot in the industry. Many times a problem arises that is local to a small line, so the overall grid would not be impacted even if the line was overloaded. In these situations, which are not uncommon, the best outcome is to be ready to act if the contingency occurs. So say a small line was overloaded by the contingent loss of a large transmission line, but the only solution precontingency was to shed load at the substation (reconfiguration of the grid would not work). As long as an operator could shed the load quickly if the event occurred, then there would be no sense in shedding the load in preparation for something that most of the time isn't going to happen.

I discussed the sagging of transmission conductors in chapter 2 under the subhead "Maintenance." What can come as a surprise is the amount of sag during heavy loads. Just to put this concept into perspective, I'll tell you about a time-speeded video I saw of a transmission line going from low load to high load and the conductors in the middle of the span sagging an additional twelve feet! If overloads occur, this is why sagging can be so dangerous. Any time a line is overloaded, the sag must be accounted for such that no danger is imposed on the public; otherwise, the line must be tripped off (and most likely load must be shed). Remember that these lines cross probably every superhighway in the country. During actual overloads, the option to do nothing is usually not available. But I can remember instances when we at Dominion had people in the field watching a particular line segment to make sure it would not sag more than expected and that we could get through an especially hot afternoon.

On hot summer days in Virginia, afternoon thunderstorms will often roll through. We system operators would be in the control room watching the load increase for the afternoon peak and trying to figure out if we had enough generation to get through the day. Around 4:00 in the afternoon, I'd start getting really nervous about the situation, especially if our generation fleet was maxed out. The peak demand for load may not occur until around 6:00 p.m. But an afternoon thunderstorm can roll through our territory and reduce the load demand by 1000 megawatts or more in a matter of minutes. Many times the storm would come through around 4:30 or 5:00 and save us from having to take further action. There were times when the operators would be praying for a storm. "Come on, thunderstorm!" they would say.

Of course, the do-nothing option is what can and does lead to a blackout on grids. NERC has defined transmission elements that are critical to the continued functioning of the grid and has set rules for taking precontingency action when a contingency shows a problem on these elements. So if a transmission operator has one of those predefined elements on his or her system, then he or she must take action within a defined time frame after the study results are known or else face significant fines. This action may include precontingency load shedding in some cases, but that is better than starting a blackout.

Information Available for the Operator

Now that I've gone through a discussion of many of the tools available to operators, I want to be clear about the information available for the operators. Operators have lots of information coming to them all the time. There are many different jobs that are covered by the term *operator*, so I will stick to the basic functions of the balancing authority (BA) and transmission operator (TOP) as defined by NERC. In most cases, these two functional operations go hand in hand. For the purposes of this discussion, let's assume that both are being performed in the same control room. I've described the BA but not the TOP. The TOP is responsible for reliability in real time of the transmission system within a defined boundary. The boundary is often the same as the BA, but not always. To be clear, I'm not discussing a generation operator, which would produce a completely different set of information.

First of all, the system that is used to gather data, perform calculations on that data, and present outputs to the operators is called the *energy management system*, or EMS. The EMS is the brain behind a complex system of communication links to substations, neighboring control rooms, NERC, and wherever else local management has decided to get data from or send data to. The EMS is a computer system that operators are very reliant on, and there are backup plans for when it fails, as everything does eventually. The basic data that come to the EMS are current flows and voltages at various points on the local grid, transmission equipment status, generation outputs, reactive power flows, and a myriad of other details. Many control centers even have some level of physical security monitoring (we did; in many substations we could actually watch who was in the substation in real time).

Most control rooms that I have visited also have a map-board representation of their local grid. The map board could have some of the information on it directly, such as voltage or megawatt flows; readouts would be LEDs. Some map boards even had the circuit breakers represented along with real-time status (open or closed). The main reason to have a map board is to provide a more global view of the grid than a computer screen can provide.

I'll give an example. Our map board was more or less geographic; that is, the top left of the board was represented by the northwestern part of our system, and the eastern part of the system was represented to the right. On Dominion Virginia Power's map board, all of the company's transmission lines are represented. I remember one summer day when we lost two 230 kV transmission lines simultaneously. When a line trips off, lights on the map board along that line immediately begin flashing to alert the operator to the change of status. Now we had two lines side by side on the map board flashing. On the computer, there were lists of breakers that were open, alarms to indicate the unplanned event, and of course changes in current flows all over. But the map board instantly gave the operator the information that the lines had tripped. In fact, one of the operators within a matter of seconds said, "Those two lines cross the James River on a single set of towers." Later we found out that a barge had been heading down river with a crane boom still up, dragging right through our lines.

Since I've brought up the issue of alarms to alert operators, let me discuss them some more. Alarms are very important to the operation of an electric utility. They can come from the EMS, the map board, a security system, or the local fire-protection equipment. The EMS can be set up to alarm almost anything: low voltage or high voltage at any number of points on the grid; single-contingency violations; current flows above a certain percentage of the line capacity; low air pressure on a circuit breaker; etc. There are far too many alarms available in the control room, so there has been a lot of work done to limit alarms to just the ones that the operator *must* have. We actually had some of the equipment alarms go directly to the support personnel who work on the respective equipment.

Some of the data received are continuously recorded on chart recorders that graph the trends of several of the important characteristics of the system. For example, there is a chart recorder in most control rooms that continuously records grid frequency. In addition to frequency, there will be the area control error that a BA uses to control total load in the BA's area, net actual interchange, perhaps flows on individual tie-lines, and many others. Most BA or TOP control centers will have completely electronic chart recorders today, rather than the old-style paper recorders. EMS systems can also provide trend charts on any data available.

One area of data available that I haven't mentioned yet is forward-looking data. This would generally include a load forecast of at least one week ahead. In addition, there would be a generation plan for the week and an outage plan. All of these data would be used in a day-ahead and week-ahead look at single contingencies to make sure that the plan doesn't have inherent problems.

Another important source of information available to the operator is information shared across the interconnection. This includes neighboring systems' outages and plans for the upcoming week so that this information can be included in the models needed to produce forward-looking contingency analyses. It would also include the information needed to identify transactions that are causing flows on each TOP's flowgates. Remember that a flowgate is just a limiting line or device that is identified to monitor the source of these flows.

The last important piece of the information puzzle is a neat system that allows for entities responsible for reliability to put information out to similar players. For example, if a BA was facing an upcoming peak for which the operators did not have adequate generation, this system could be used to put the message out to all balancing authorities, transmission operators, reliability coordinators, et al., to say that the BA needed some help. In fact, it is expected that any system that is projecting potential trouble these days will let all the other players know so they can help, or at least be aware of the risk.

I haven't even mentioned the phone, which is not high-tech but which is in use much of the time. Many times I would walk into the control center to find that everyone on shift was on the phone. This is due in large part to the communications with field personnel. For example, as field people are ready to perform some switching of electrical equipment on the grid, they must make contact with the system operator for final permission. There are also constant communications with neighboring utilities, as well as with the folks who handle the distribution system. All phone conversations are recorded. Occasionally, if there is some dispute about what was agreed upon, we even go back and listen to the recording.

In the last two chapters I have described many of the activities that grid operators and engineers do to keep the grid as reliable as possible. In the next chapter I am going to delve into the governmental regulation of the electric business with a focus on how that regulation impacts the design and operation of the grid.

CHAPTER 4

THE GOVERNMENT'S ROLE IN THE GRID

My lovely wife and I were invited to an industry conference on competition in the electricity business. In the evening the conference sponsored a wine-tasting event followed by an open microphone for making comments about competition. My wife is someone who does not hold alcohol well. She somehow managed to get the microphone and said she was worried that with a competitive electric business there would no longer be a clear line of responsibility if the lights went out. "Who would I call: a marketer, the local utility, or another utility from a different state? I mean, I sleep with the guy who runs the transmission grid, but I couldn't get my power back for seven days after the last hurricane. What will I have to do after competition?" I hid under the table.

Introduction

In this chapter, I'm going to explain the roles that federal and state regulators play in relation to the grid. I will also relate these roles to those of some of the other, quasi-government players in the industry. The regulators are extremely important, as decisions they make can have a huge impact on grid reliability. I am not a lawyer, so my discussions will mainly focus on impacts to the operations and reliability of the grid rather than on legal arguments.

William L. Thompson

Federal Energy Regulatory Commission (FERC)

Of all the regulators, FERC has the biggest influence on the eastern and western grids, since, as I mentioned before, transactions on those grids are considered interstate commerce and fall within federal jurisdiction. FERC's role in Texas is somewhat diminished because of the intrastate scope of the grid there. States play an important role as well; I'll get into that topic later. NERC is the main quasi-government player with respect to standards development, but NERC must rely on FERC's authority to enforce these standards.

FERC is run by (up to) five commissioners, who are appointed by the US president and approved by the Senate. Commissioners typically have a legal background with experience in regulation. The seated commissioners serve five-year terms. After a presidential election, they are not immediately replaced, although over the past twenty years there have been a few cases where a commissioner has resigned after an election. There are often vacancies on the commission, so don't be surprised to see only four or even three commissioners at any given time. And there is a statutory limit of three seats for a political party. The United States president appoints the chair. FERC's decisions are not subject to congressional review/approval. However, Congress can and does define the role of FERC via statute, as I will discuss below.

FERC's responsibility in the electricity sector is to regulate wholesale sales of electricity as well as the transmission of electricity for interstate transactions. FERC is also responsible for enforcing mandatory requirements (standards) for the reliability of the grid. In addition, FERC must approve any mergers for jurisdictional companies, and can grant or deny the right to charge market-based rates (as opposed to a cost-based rate) for wholesale transactions to those entities that can be shown not to have market power. FERC is responsible for many things other than this, but these are the areas of focus to examine how the commission impacts the grid.

I will provide an overview of the most critical laws of Congress and the rulings FERC has issued over the past twenty-plus years. These are the rulings that have guided the ownership structure and much of the operations related to the grid today. They are as follows:

- the Energy Policy Act of 1992
- FERC Orders 888 and 889
- the Energy Policy Act of 2005
- FERC Order 693 and additional orders on standards

The way to understand these laws and orders (and many others I won't get into) is to think in terms of the policies that are driving them. In general, the policies include energy conservation, just and reasonable rates for all transmission users, a level playing field for all players in the electricity wholesale business, a strong national energy infrastructure including a reliable grid, and wholesale competition including market-based wholesale rates for both energy and generation capacity. This may sound like a fairly simple list of policies, but the road to get there is quite a story. I will start with the Energy Policy Act of 1992, discussing how that led to some major changes in how the grid is operated.

The Energy Policy Act of 1992 (EPAct)

This act, often called EPAct, was passed to enhance competition in the electric business. It provided for a new class of generator owner (called an *exempt wholesale generator*) that could be free of many of the requirements placed on electric utilities. But perhaps even more important, EPAct made it clear that the transmission owners are required to provide service to these new, competitive entities. In other words, the transmission business has to operate as a common carrier for electric service; no longer can the transmission owner give priority or preferential treatment to an affiliated generator.

Following is a simple example of this issue. Imagine that, prior to EPAct, the ABC generation company wanted to build a power plant within the boundaries of transmission owner TAB. Also assume for now that the owner of the transmission in the area, TAB, was a vertically integrated utility with no statutory requirement to buy the output of the new plant. ABC wants to sell the output to TAB, but the negotiations break down when it becomes clear that TAB won't pay enough for ABC to make an adequate return on the investment. So now ABC wants TAB to "wheel" the power to a neighboring utility, that is, to transmit the power over TAB's transmission system to another entity. There is no requirement that TAB should do this for ABC, so TAB determines that there is no transmission capacity available to wheel the power. ABC is stuck with the low offer made by TAB if they still want to build the plant (of course, with this playing field, the plant would never be built).

I would like to add that prior to EPAct there was a law passed that would provide for certain plants to sell to the local utility. The Public Utility Regulatory Policies Act of 1978 (PURPA) would have provided for the statutory requirement for the local utility to purchase the output (at "avoided cost") if the plant met certain criteria such as cogeneration, where steam or heat was sold as a by-product of generating electricity. PURPA did not go so far as to require wheeling the power across the grid, however.

Now EPAct comes along in 1992 and says that FERC should require all transmission owners to provide open access to the transmission grid for all generating companies. Once it became clear to those of us in grid operations what this act mandated, we realized that wholesale competition was coming soon. But to be honest, we had no idea of the complete ramifications of EPAct.

FERC Orders 888 and 889

After a few years of intense debate, FERC, in April 1996, finally issued the landmark orders that changed everything in the grid business: Orders 888 and 889. These orders were called, simply, "Promoting Wholesale Competition through Open Access Non-discriminatory Transmission Service by Public Utilities" and "Recovery of Stranded Costs by Public Utilities and Transmitting Utilities." These orders actually implemented the changes that EPAct initiated. I'm going to summarize the biggest impact areas that these orders had on the grid. Areas I'll define are the following:

- open access to transmission
- generation transactions

- separation of affiliated businesses
- unintended consequences

Open Access to Transmission

The orders required transmission owners to develop and file with FERC an Open Access Transmission Tariff that defined conditions and costs for a number of services for the transmission business. In fact, the orders actually included a fully written pro forma tariff that transmission owners were essentially told not to change. All the rates charged and the conditions for service were required to be approved by FERC. In addition, the orders required that the rates and terms should be posted in real time, for all to see, and that the affiliated generation had to give no preferential treatment for any services.

The site for the postings was called the Open Access Same-Time Information System (OASIS—an acronym proposed by a Dominion employee). Not only was posting our prices for various transmission services required, but also we were required to post the availability of transmission service on an hourly, daily, and monthly basis. Requirements for OASIS were quite specific and very complex for how postings were to be made, how they could be changed, and how no information could be transferred to an affiliate any sooner than it was available to all competitors.

The policy goal here was to augment a competitive market at the wholesale level in electricity. The one big step in getting there was to end preferential treatment for generators affiliated with transmission ownership, thereby lowering the hurdle for new entrants who wanted to compete. Prior to these orders, transmission owners had a number of ways to exclude new entrants and to favor their own affiliates. Among these was the practice of denying access to transmission for new entrants on the basis of reliability. Now that the operation of analyzing transmission capability was transparent, any available transmission capacity had to be offered to all entities at the same time and at the same price. In addition, any information about the transmission grid had to be available to all players at the same time.

Generation Transactions

Up to this time, vertically integrated utilities had developed "interconnection agreements" with their neighbors that provided for fairly simple arrangements to conduct transfers of energy and/or capacity. These agreements had been approved by FERC years ago, and utilities had operated under them for at least twenty years. Interconnection agreements generally treated transactions as "bundled" deals, that is, the generation of energy and the transmission of the energy was bundled into one package deal. EPAct and the subsequent Orders 888 and 889 forced utilities to unbundle transmission from generation (see below). Existing contracts were grandfathered and were therefore allowed to continue as is. Although it has been over twenty years since EPAct was passed, there are still bundled contracts being used as I write this.

After these orders were issued, transactions on the grid changed dramatically. Where a bundled deal before the orders could simply deliver energy to the border of a utility, after the orders this transaction would be in two parts: a transmission-path purchase and a separate contract for the energy. But to add a level of complexity, the purchaser of the transmission path could buy the path across several systems in order to transmit energy a very long distance. The interconnection agreements did not allow for this, as they were made only between two interconnected neighbors. Prior to the orders, to move power from Georgia to Virginia would require each transmission owner on the path to buy the energy at one border and resell it at another border (the border I am referring to is not the state border, but the border of ownership of transmission). After the orders, the transmission owners were not allowed to buy the energy, as that would have violated the requirement for separation of the transmission from the generation business. The orders made it easier for a generation company to do transactions over large distances, even though there was the added complexity of the deal being done in two parts. This is the reason why I say that the restructuring of the business on the grid improved our ability to maintain a reliable system (after all the bugs were worked out, of course).

Separation of Affiliated Businesses

A large part of the orders focused on the granting of preferential treatment to affiliated generation. Since many of the utilities that these orders applied to were vertically integrated, it was easy to see why this was such a big concern. FERC was not going to get true competition if the affiliate generator could simply have someone walk into the control center and see what was going on. That would be like playing football against a team that had the right to know which plays were going to be run. The interface between affiliate generators and the transmission operators was only allowed through OASIS, a truly transparent system.

Following is an oversimplified example to show why this separation is such a big deal. Say for example that a major transmission line trips out that usually carries power into a big city. Imagine also that the only remaining electrical path into that city is now through transmission owner TAB. TAB contacts its affiliate generation operators and tells them of the tremendous advantage they now have to sell power to the city. Their affiliates buy up all the available transmission capacity to the city before any competitors even realize what has happened.

The outcome is intended to be a level playing field for all electricity generation at the wholesale level. Transmission is operated in such a manner that all generation players get access to the same information at the same time. If an affiliate so much as asks for a favor from a transmission operator, management becomes very concerned. The price for wheeling power through a transmission owner is always the same for everyone in the generation business—no exceptions.

Unintended Consequences

I have already discussed some of the consequences of these orders in chapter 2. None of us system operators were ready for the volume of these transactions on the grid. Keeping track of everything from a big-picture perspective so that it could be unwound when things went awry was impossible at first, before the tools were built. There were almost certainly days when the grid was one contingency away from a major outage. In those days there were very few of us running an online single-contingency analysis; and even if we were, there was little that could be done. I don't know how many times an operator somewhere was close to shedding customer load due to "loop flows" from unknown sources going through his or her system.

It was understandable, as in the example above, that the orders did not allow discussions between transmission operators and generation operators. However, there was actually no restriction on information about the generator getting to the transmission provider. In fact, the grid is extremely hard to operate and control without information. The problem was that after FERC had levied some hefty fines for affiliates who broke the rules, some of the generation companies reacted by telling their operators not to talk with the transmission operators. One day an operator at a nuclear plant, when asked when he expected to return the unit to service, told the transmission operator that providing this information was not allowed. It took years for all the players to get straight what was allowed and what was outlawed.

FERC did allow for a company to declare an emergency and then converse freely with all parties until the problem was resolved. The caveat here was that the transmission entity was required to keep a log of all conversations that otherwise would have been illegal. We had to use this allowance when a hurricane (Isabel) came through our territory in 2003. Frankly, it was the only way we could have gotten through the storm. We had lost more than sixty transmission lines, several generators, and around 80 percent of the load. We needed good information (and, of course, generation) to keep the rest of the system energized.

Another serious consequence that I touched on earlier was the direction of capital investment for the industry. Some utilities decided that investments in transmission and distribution were not in the best interest of their stockholders. It appeared that a greater return on investment was available in generation. Consequently, for several years capital investment was directed toward generation rather than toward the grid. As if this wasn't bad enough, some utilities decided that maintaining the transmission grid just didn't pay. Tree trimming, probably the biggest maintenance cost related to the grid, was minimized by some transmission owners until the aftermath of the northeast blackout of 2003 clearly showed what was going on.

But the biggest and most surprising consequence came out of the fact that the FERC rules issued via Orders 888 and 889 were legally required, but all of the operational rules that ensured reliability were not legally binding. So after those of us in the industry had finally gotten our act together to deal with the transactions, the loop flows, and business separation, we found that some grid operators felt that following the reliability rules was optional. This created some incredible situations that to this day I find amazing.

The worst case was a hot day in the late 1990s when energy was selling for $7,500 per Mwh, if you could find it. Companies were willing to pay that much for the energy if the result was that they wouldn't need to shed customer load. The frequency in the eastern interconnection was low, and we all knew that things were on the edge. It turned out that a major player in the Midwest had somehow failed to find enough energy to serve their load. In fact, as I recall, they were short for several hours, by as much as 1600 Mw. What did they do about it? Nothing. What was their penalty for putting the grid at risk? No penalty.

What should they have done? The rules say that in this condition, after trying all other options to buy emergency energy from others or to reduce load through public appeals, the transmission operator should have implemented a load shed in his or her own territory to an extent that would balance load to resources. What this company did is what operators call "dragging on the system." Essentially the company just decided to lean on all the rest of us to provide the extra power they needed. This extra power came from two sources: (1) the frequency bias setting that I discussed in detail in chapter 2, which would have all the other BAs providing an extra boost to the grid due to the low frequency, and (2) the slight reduction in demand throughout the entire system that is due to the lower frequency on the grid. These two factors helped keep the grid from collapsing, and amounted to the 1600 Mw that the company I mentioned was "dragging." To put some perspective on the economics of this event, one hour's worth of 1600 Mw (1600 Mwh of energy) during this time was worth around $12 million.

William L. Thompson

The Energy Policy Act of 2005 (EPAct05)

The event described above and many others like it led many of us in the industry to clamor for reliability rules that were mandatory, with fines to ensure that companies didn't simply decide to pay the fine rather than do the right thing. At this time, the late 1990s to the early 2000s, the FERC did not have the authority to level fines for violating reliability rules. Unfortunately, neither did NERC. The problem here gets back to the fact that those of us who operate the grid are so highly dependent on each other. A problem ignored five hundred miles away can lead to a blackout here, and that is not acceptable. Then the northeast blackout of 2003 occurred and brought all this concern to the forefront.

EPAct05 included a lot of important things that I won't get into. But stuck somewhere deep in the law was a provision that gave FERC the authority to fine entities for not following the reliability rules. And the fines could be up to $1 million per day for noncompliance (more on this later). In order to implement this provision, FERC went through a process that certified NERC as the Electric Reliability Organization (ERO), tasking NERC with developing the standards that would be enforced. The standards that we had been using in the industry prior to this were not precise enough to be interpreted legally, so a huge project was started up to rewrite standards. This process continues to this day.

I want to consider the impact this law had on those of us who operate the system. On the positive side, we now had more than peer pressure to leverage when trying to get our neighbors to do the right thing. Over time this has had a very positive impact on the reliability of the grid. On the negative side, though, was a new focus on many aspects of our standards that were not directly related to reliability. For example, there is a provision of the standards that says that every protective relay must be maintained on the schedule committed to and that maintenance has to be documented (if it is not documented, it didn't happen). Is it possible that the grid operators would become so focused on documentation that they would fail to act in order to avoid some big event? The other issue has been the additional staff in the back office that now have to spend all their time on documenting compliance with the standards.

To help in explaining the new authority granted to FERC, I will focus a bit on how the fines work. The provision of $1 million per day was a little vague to some of us at first, as we thought about noncompliance as a onetime event. That is not the right way to look at it, though. The right way to look at it is expressed in the following example. Say that as part of the standards a transmission owner has developed a commitment to trim trees to a twelve-foot minimum clearance from the conductors. Say also that someone finds a tree growing into that zone. Furthermore, it is somehow determined that the tree has grown two feet into that clearance zone. FERC's thinking on this type of situation is that the noncompliance began when the tree first grew into the zone. If the right-of-way is for a line very important to grid reliability, then the fine could be $1 million multiplied by the number of days this species of tree takes to grow two feet. Such a fine will definitely get someone's attention. Fortunately, there are many mitigating factors that can be applied, so fines for failing to trim a tree have been in the range of $100,000 to $250,000. These are expensive trees!

There was one event that I will point out that really sent a shock wave through the industry. One day in 2008 outside of Miami, a Florida Power and Light (FP&L) technician was doing some work in a substation and made some really bad procedural mistakes. What he did was turn off some of the relay protection equipment so that when a device failed, there was nothing to remove the fault from the grid. The result was a huge electrical perturbation throughout the entire eastern interconnection. In Florida, there was a loss of about 4000 Mw of load, which took several hours to recover. After an extensive investigation, FERC fined FP&L $25 million for noncompliance. The strange thing here was that in the final order from FERC, there was no specific standard mentioned that FP&L had failed to follow. I guess the message was simply that if a company is going to put the grid at risk, it is going to pay a big fine.

FERC Order 693 and Additional Orders on Standards

In 2006, NERC filed a large collection of standards with FERC for the latter's approval. NERC needed a formal FERC approval in order to have the legal right to implement the mandatory compliance program.

FERC issued its order as FERC Order 693 in March 2007, which I have highlighted because it was the first step in mandatory compliance to standards. The order, 517 pages in length, accepts some standards into the program, denies some others, but also orders NERC to begin the process of changing many of the standards. After this order came out, we in the business realized what a long, tortuous process getting to mandatory compliance was going to be. At one time I actually thought that the whole thing could be done at NERC without federal oversight—but then I am not a lawyer, only an engineer.

The first standards were legally enforceable in June 2007. To this day, in 2015 as I write this chapter, the industry has been using a large set of standards that are partially mandatory and partially not (not approved by FERC). In addition, existing standards are being changed constantly. The mandatory treatment of standards has improved grid reliability, but the constant changes present a risk. The operators of the grid have to remember and comply with all these rules in real time, day in and day out. Operating the system was complicated before this added focus on compliance. There were times when we system operators just kept saying to ourselves, "It is getting better!"

Another important point to mention here is that FERC doesn't have a legal right (yet) to issue the standards itself. FERC has to go through the process of having NERC develop a proposed standard and then submit it to FERC. FERC can accept the proposed standard as is, accept it with some changes, or deny it. What they can't do is issue a mandatory standard for the industry.

One Critical Federal Issue: The Independent System Operator

There is an important issue that I want to delve into, and that is the issue of having an independent entity become the operator of the transmission grid. The concept of "independent" means that the operator is not affiliated with the ownership of any generation. For a time there was a clamoring from the market side of the industry for FERC to order all transmission to be operated by independent entities. The belief was that this would be the only way to ensure that the transmission operator did not give preferential treatment to its affiliated generator.

In the late 1990s and into the 2000s, there was an incessant push for this independence, especially as examples of abuse seemed to routinely pop up. In Virginia, state legislation in the 1990s required the use of an independent system operator (ISO). It seemed very likely that FERC would issue an order requiring an ISO for all transmission; we waited to see what FERC would say. Interestingly, though, FERC never issued the order. The thinking was that they did not have the authority to order the transition to ISOs, although Congress could have included that in EPAct05 if the politics had allowed it. Frankly, there were many in the industry who felt that the transition to ISOs was not necessary and would be detrimental for a number of reasons. There were several states that were opposed to the concept, especially those south of Virginia.

Geographically, those transmission-owning companies north of Virginia decided to join an ISO, while those south of Virginia did not. There are three major ISOs in the eastern interconnection: ISO New England, New York ISO, and PJM (*PJM* used to stand for the "Pennsylvania, New Jersey, and Maryland" pool, but now it is just a name, as I will explain below). In chapter 2 I brought up the issue of reliability coordinators (RCs), which were instituted in order to have a broader view of the transmission grid. But don't confuse these two issues; the ISO is independent from generation and handles a market without preference to any players, whereas the RC is not necessarily independent but does have authority to order curtailment of transactions under some circumstances. The ISOs I mention are also acting as RCs, but all the RCs are not independent entities.

Since Virginia law ordered Dominion Virginia Power to join an ISO, we eventually joined PJM. This means that we agreed to turn over to PJM the control of aspects of transmission operation that could be used to give preference to a generation affiliate. So in 2005 we had a turnover process to PJM operations. PJM is a nonprofit entity that does not own any generation. After several transmission owners (IOUs) decided to join PJM, including American Electric Power (AEP) in Columbus, Ohio, and Commonwealth Edison in Chicago, Illinois, the initialism PJM didn't make sense. After some investigation into a name change, the organization decided to retain the PJM moniker without its actually being an initialism for anything.

What does PJM control? PJM operates a transparent market for wholesale energy, which means that generators bid their prices to PJM. PJM then determines the mix of generators that will run. But as I have mentioned in previous chapters, there are constraints to delivery on the grid. So PJM has to determine not the lowest-cost mix of generators to supply demand, but the lowest-cost mix that is reliable. This would inevitably result in some generators running that are not the least-cost bidders. Why? Remember from chapter 3 that to relieve overloads on the grid the operator must reduce generation on one side of the constraint and raise generation on the other (redispatch). To implement this, a cost higher than the least cost must be paid for the generation on the load side of the constraint.

PJM accomplishes the equivalent of redispatch via a methodology called *locational marginal pricing*, or LMP. This fairly complicated system pays more for desirable generation on the load side of the constraint and pays less for the generation on the other side of the constraint. Doing this relieves the overload, but it adds energy costs. With this type of control on the generation market, it is understandable why many on the marketing side think that only an independent entity should do this.

There are other aspects of the business that ISOs generally control. The first I'll discuss is the market for generation capacity. As mentioned above, the real-time energy market is controlled by an independent entity. It became apparent in the late 1990s that an energy-only market was not going to work over the long haul. Generators that are competing in a market with excess generation will simply close up and send everyone home if they can't make a return on investment from their profits on energy sales. If that is allowed to happen when the peaks in demand happen, as they always do, there won't be enough generation to serve the demand. This may be okay in a lot of commodity markets, such as oranges, but in this country shortages of energy are not acceptable. Additionally, there was a serious concern that no one would build new machines capable of generation since the risk was too high that the energy market alone would not support the investment. So generators are also paid for their available capacity, whether or not they actually run.

Another issue that these independent entities control is the sale of available transmission capacity. It has the authority to analyze the transmission system and determine how much additional transmission capacity is available as well as the methodology for selling that capacity on a completely transparent basis. Remember the OASIS system that I mentioned earlier? When Dominion Virginia Power joined PJM, we turned that entire process over to them.

It was never tested whether FERC had the authority to order ISOs throughout the grid, but there was a time when the threat was very real and had an impact. This was especially true for those transmission-owning companies that tried to merge during the late 1990s and beyond. To do a merger or an acquisition requires FERC approval, and FERC simply held the concept of ISOs over the heads of the companies that were attempting to merge. As it turned out, however, the winds changed during the 2000s, and most companies in the Southeast never turned over their transmission to an independent operator. To put some perspective on this, approximately two-thirds of the load in the United States is within the purview of an ISO.

So what is left for the transmission owner's operators to do after the company joins an independent operator? Actually there is still a lot left in the hands of the member companies. First of all, although I've said that the ISOs have "control," I have really used this term loosely, as all actual controls for the grid remain with the transmission owner. The ISO would establish all the market rules, as I've mentioned above, and hopefully a reliable system will be the outcome. However, if there is ever a need for opening a transmission line or for shedding load, the transmission-owning companies will have to implement the order. Also, many "local" flow problems that are not controllable using redispatch (and there are a lot) are simply put into the hands of the owners. For example, local flow problems that would require a load shed after a single contingency, discussed in chapter 3, are turned over to the transmission owner. In addition to this, though, is the issue of maintenance of the grid, which is in the hands of the owners. Maintenance usually requires outages. These outages have to be approved by the ISO but are completely performed by the owners.

Department of Energy

The Department of Energy's (DOE) stated mission, taken from their website, is "to ensure America's security and prosperity by addressing its energy, environmental and nuclear challenges through transformative science and technology solutions." I didn't have a lot of interface with the DOE in my time in the business, but I would be remiss if I did not mention them. They got heavily involved in the 2003 blackout analysis, taking the lead on the United States–Canada Power System Outage Task Force. For a time, it was unclear to me whether it would be DOE or FERC that was going to enforce mandatory compliance to the NERC standards. Anyway, as it turned out, FERC does the enforcing. I think of DOE as being more technically based than as being an entity that gets into the details of ensuring compliance with rules.

The DOE is a great source of data on many aspects of the energy business. If someone wants to know the amount of generation due to nuclear energy in 2008 in the United States, all she or he must do is take a look at the Energy Information Agency (EIA) website. EIA is a division of DOE. This is one reason why I try not to dwell on data/statistics in this book, since they are so easily available. In fact, most of the data in the next chapter on generation are adapted from EIA data.

State Regulation

The states are responsible for regulating the retail delivery aspects of the electric business, including the rates paid by final-use customers served by IOUs. Cooperatives and municipalities are usually regulated by their own board or city government. They are also responsible for all siting issues related to power lines and generators (with one possible exception that I'll mention below). Some states get involved in generation reserve, or requirements for renewables. In this section I will discuss in more detail the following topics:

- electric delivery reliability
- siting of electric lines and substations, including transmission
- a failed attempt at federal jurisdiction in siting of transmission lines

- general principles of rate making
- a national association for state regulators
- a typical residential customer's bill

The states, of course, are quite interested in electric delivery reliability, although their specific jurisdiction does not include the transmission grid, since FERC has jurisdiction in most states. That said, however, state regulators can make life extremely difficult if the overall reliability of service is not up to par, regardless of whether the problems are with the distribution system or the transmission system. State regulation can certainly lead to specific targets for reliability and can also stipulate the type and amount of generation within the state.

Siting of transmission lines is within state, not federal, jurisdiction (in fact, smaller lines within a single county may not even be required to get to the state level for approval). An electric utility would make a case for the need for a new line, propose a route, maybe even offer some optional routes, and file this with the state regulatory body that oversees the electric business. In the last couple of decades it has been difficult to get a new line built. Nobody wants a transmission line in his or her backyard. But as demand for electricity increases, the need for building major transmission lines steadily increases.

Utilities generally encounter a huge controversy around any proposed new transmission line. The state regulators are not always in a great position to fully comprehend the need for a line, as at times the need is based on large regional (i.e., multistate) studies. There have been cases where the state regulators took years to make a decision. That is why in the EPAct05 legislation there is a section that allows for FERC to step in and make the decision, thus taking siting jurisdiction away from a state. The conditions in which this could happen are narrowly defined, include only a very few select corridors defined by the Department of Energy (National Interest Electric Transmission Corridor), and can only be invoked if the state has not acted within a year after the initial filing. In the years since this order came out, the United States Court of Appeals for the Fourth Circuit has ruled that FERC can only act if the state has not. FERC cannot reverse a denial by a state. As far as I know, this provision has had zero effect. It doesn't look like it will be used anytime soon, either.

As I mentioned above, the states are tasked with establishing the rules under which utilities serve retail load. The push for complete retail competition seriously slowed after the California debacle, but it seems that all states have different approaches to the regulation of the business. In a nutshell, the utility is granted a rate of return on assets based on filings before the regulatory agency. From a rate-making perspective, many of the costs that the utility incurs for the bundled service of producing and delivering electricity to the customer are pass-through costs. This means that the utility can collect from customers only the actual cost required to perform the service. Pass-through costs include fuel and maintenance, for example. Then how can a rate for each kWh be set?

Typically the IOU files with the state a rate case that includes the entire expected pass-through costs for a period of time, along with a detailed analysis of a return on assets that is supported as fair and necessary for the company to remain as a healthy entity. An estimate is made of the total energy delivered for a period of time (e.g., one year) and of all the direct costs associated with that level of delivery. The IOU can then tack on the rate of return on assets, along with the total value of assets, and come up with a total revenue requirement. The total (annual) revenue requirement is divided by the total expected delivery to arrive at an average cost per kWh. All of this information in the filing is subject to review by the state regulators and can be contested (and usually is). This explanation is oversimplified, as in reality there are lots of other issues, such as that there may be a different rate per kilowatt-hour for each class of customer (commercial customers may pay a higher rate than residential, while industrials pay less). Also, since the going-in estimates for fuel cost and total energy served are going to be wrong (nobody is perfect), there are provisions for subsequent corrections of these pass-through costs.

For example, if the weather for the year turns out to be severe, demand for electricity will be higher than expected and the utility will spend more on fuel than estimated. But also, the utility will collect more from customers (since the rate per kWh was set to recover the allowed return for a lower number of kWh delivered) and therefore on paper will make a higher return on assets than allowed. These overearnings, plus interest, would then be resolved in future years by a reduction in the rate.

What about transmission expenses, since the FERC is responsible for those rates? The filing that the integrated utility makes with the state would have to show the transmission rates that were approved by FERC, where the utility has taken a similar approach to averaging a cost per expected kWh delivered. The state may not like the rate approved by FERC and may attempt to exact some reduction in transmission costs via a reduction in return on assets for what is in the state's jurisdiction. This, of course, will lead to the lawyers making a fortune. (I made up this scenario just to show how complex this can get.)

Rate making can be complicated. There are many people within each utility and in each state who spend their lives working on it. There are many other factors in rate making, such as whether industrial facilities can get competitive energy off the grid (for example, can they elect to wheel power from another utility across the grid?), the difference between commercial rates and industrial rates, federal government establishments and rural cooperatives (they are not in state jurisdiction), how often a utility must refile for a new rate, allowance for funds used during construction, and commercial rates that subsidize the cost of residential rates. Frankly, a book on this subject alone would be necessary to get into details. I don't want to get too deep into that aspect of the business, but I do think it useful to present some basic thoughts.

The state regulators have an association that allows them to compare notes on the regulatory models. This association is called the National Association of Regulatory Utility Commissioners (NARUC). I presented at a NARUC meeting (about our 1994 load-shed event) but was never allowed to sit through one. I suspect that there must be some interesting discourse about the pros and cons of many facets of rate making. The challenge is finding the right mix of a reasonable price of electricity for the customers and a rate of return for the utility that provides incentives to build and maintain a reliable system for the present and for the future. The reason I say this is that too low a rate of return may be politically popular in the short term but could lead to higher costs to customers in the long term, due to decay in the infrastructure. Why? Because if not incented to build new infrastructure, given that the cost of capital is higher than the allowed rate of return, the utility will not build it. The decay will show up in lower reliability and higher costs for energy.

I want to take a quick look now at the rates paid by a residential customer from his or her perspective. That customer gets a bill that will show total kilowatt-hours used for the month. The costs for that usage could be broken down in several ways but most likely will not have a demand charge (a charge based on the maximum kilowatt demand for the month). I would say that most of us pay a rate per kilowatt-hour regardless of time of usage. In some cases, the bill might show a time-of-day differential, if the customer has the up-to-date meter to support that level of sophistication. There may be a supplemental cost in the summer to help pay for the peak usage, or perhaps even a credit for participation in a demand-reduction program. There may also be a line item for the transmission cost.

One point I want to leave the reader with here is that the wholesale market and the actual price paid at the retail level by residential customers are not directly linked. For example, the wholesale market on a hot day after several major units trip can go sky-high, but the rate a residential customer is paying for that energy does not change. This is one important reason why some early programs to deregulate the electricity market failed, as there was no incentive at the retail level to reduce usage on those expensive days.

North American Electric Reliability Corporation (NERC)

NERC has been mentioned several times so far. It is now time to define their role. NERC is an international nonprofit organization that establishes the rules by which we design, maintain, and operate the grid. After a blackout in the northeast in 1965, NERC was formed to establish reliability rules and to show the United States federal and Canadian regulators that the utilities can control their own. NERC was originally a council made up of representatives from the utilities. They established rules for various aspects of grid reliability, including design, communications, maintenance, and operations. The only method of enforcement of these rules was peer pressure, which worked fairly well for many years.

EPAct05 established the concept of an Electric Reliability Organization (ERO). Rather than give FERC the role directly, Congress decided to have an industry group establish the rules under the authority of FERC. NERC applied to FERC to be the ERO, and FERC granted NERC that role in 2006. Of course there was coordination with equivalent regulators in Canada and Mexico as this evolved. Today, NERC is a corporation rather than a council. They are independent of ownership by any of the players in the industry. Also very important is the fact that NERC committees have representatives from all players in the industry, including many marketers. NERC's website says that they serve more than 334 million people in North America.

The challenge at NERC has been converting old standards that were developed with peer-pressure monitoring in mind into new standards that were going to be legally binding. This process continues today. Just to give an example, the old standards required that a system operator be knowledgeable about the protective relaying equipment. What does it mean to be "knowledgeable" about relaying? Is there a test? (No, there isn't.) We all thought we knew what this statement meant in the old days, but how can compliance be monitored with statements like this one when the fines for noncompliance can be enormous? Another role of the ERO is to monitor compliance, and this has also been a tough assignment. NERC depends on eight regional entities to perform most of the monitoring. These regional entities are also independent corporations.

NERC's home office is in Atlanta. A lot of the work NERC does is performed by committees made up of industry representatives. The process of standards development is slow and tedious, as competing issues and interpretations always seem to come up. Although there are some emergency processes that can bypass normal due process, the work to change, add, or delete a standard is daunting. And ultimately FERC (and the Canadian equivalent) has to approve all standards. In spite of all this, there have been some huge improvements in grid reliability since the standards became mandatory in June 2007. But these strides came with a lot of work, frustration, and false starts.

Other Entities That Impact the Grid

Now that I've gone through the most important entities for the grid, I'll mention a few more that have a significant impact on how the grid is designed and operated. In this section, I will discuss the following:

- the North American Transmission Forum
- the Institute of Electrical and Electronics Engineers
- the Electric Power Research Institute

The North American Transmission Forum (Forum) is organized to further the drive for a more reliable grid. Membership is restricted to any entity that "owns, operates, or controls at least 50 circuit miles of integrated (network) transmission facilities," with allowances for some entities that operate but don't own transmission. In the days before wholesale competition, NERC was a great forum to air out problems, share experiences, and seek excellence. As the role of NERC transitioned to more of a compliance organization (or the "cop," as some people called them), the transmission operators felt that something was missing. They felt the need for a forum where grid reliability was the only focus and where discussions could take place that went beyond the standards and compliance but also sought improvements in design or operations. The Forum's meetings are somewhat closed and results are confidential, which enhances their ability to achieve the mission.

The Institute of Electrical and Electronics Engineers (IEEE) is a nonprofit professional association of members throughout the world. IEEE develops its own standards, which are generally accepted worldwide. Their impact on the grid is one of providing the highly technical standards for the equipment that the grid relies on every day. For example, a 500 kV circuit breaker at a substation is built to meet standards developed under the authority of the IEEE. NERC standards do not infringe on IEEE standards but actually build on those standards.

The Electric Power Research Institute (EPRI) is a nonprofit organization that conducts research in all technical areas related to the electric business. Supported in large part by the industry, the EPRI also brings outside expertise to its work. This organization has had a big impact on the grid, not only in equipment design but also in efficient control methodologies. The competitive environment we live in today has been a challenge to EPRI's longevity, but so far they continue to drive many research areas.

In this chapter I have described the governmental regulation that impacts grid development and operations. The discussion is at a fairly high level. The details of regulation in the electric business change constantly. In the next chapter I am going to shift into a discussion of the various electric generation options available, with a focus on how each option affects grid operation.

Chapter 5

GENERATION

On May 5, 1999, the front page of the *Richmond Times-Dispatch* had an article about Y2K. The article opened with, "Bill Thompson said he is not running out to buy a portable generator out of fear that the lights will go out on New Year's Day. That's good news, because Thompson is manager of power supply for Virginia Power." At that time, those of us who operated the grid were getting pretty confident that the Y2K scare was overblown. It turns out that in this case our confidence was well placed, as nothing significant happened. By the way, I have bought a portable generator since I retired. My wife made me do it!

Introduction

The subject of generation sources can be very controversial to many constituencies. Antinuclear folks don't want the risk of nuclear generation. People worried about climate change don't want any fossil (coal, oil, natural gas) generation. There is beginning to be concern about the long-term effects of our latest method to extract natural gas: fracking. There are folks who think that with conservation programs, there is no need for additional generation. Perhaps there are those who believe that with a concerted effort placed on developing renewable sources such as wind, solar, geothermal, and/or waves, the entire fleet of fossil fuel and/or nuclear plants can be retired. And, of course, there are those who are simply off the grid. What does the future hold for generation sources of our electricity supply?

In this chapter I will explain the pros and cons of different generation sources. The perspective of this assessment will be that of a system operator, the person who is expected to maintain the grid. The system operator on duty is not concerned with the politics of nuclear, coal, or renewable sources. He or she is concerned with getting through the day without a catastrophic event, such as a grid collapse. Public policy can change the generation mix that supplies electrical energy, and public policy can change how energy is used by mandating appliance efficiencies, but as I've mentioned previously, politics can't change the laws of physics. People responsible for operating the grid all share the concern that someday public policy will result in rules that are insurmountable when trying to maintain a reliable system. That is one reason why there is pushback from many operating entities about the extent of conversion of fossil fuel to renewable generation.

General Terms and Concepts

Before I get into the details of generation sources, I will discuss a few general concepts that need explanation. I will divide this discussion into economic terms and concepts, and technical terms and concepts. Within some of the generation types below, there will be a few additional concepts specific to that type.

Economic Terms and Concepts

The first set of concepts I'll discuss is related to generation economics. In this section, the following topics will be addressed:

- generation plant vs. unit
- construction cost
- levelized cost
- going-forward cost
- incremental cost

Everyone probably understands the concept of large generation plants, which convert one form of energy (such as fossil fuel) into electrical energy. These plants may consist of several units, each unit able to deliver power measured in megawatts. Although in a plant there may be some shared facilities, such as the coal pile, a unit is generally thought of as a single electrical generator and its directly associated equipment, such as a boiler. Given the economics of building and operating generation, most of the units we deal with are located at plants that have multiple units and shared services. A collection of wind turbines is called a *farm*, whereas a large solar installation could be called a *solar array*. A solar installation at a residence is called a *solar installation*. Large installations of fuel cells can be called *farms* or *stacks*. Any form of generation at residences would be called *distributed generation*. Hopefully that helps with some of the terminology.

Often the economic analysis of type of generation to build is simplified by estimating the total cost to construct divided by the total capacity in kilowatts. This number, expressed as dollars per kilowatt ($/kW), provides a yardstick for the cost of new capacity but doesn't compare the lifetime costs of two different types of generators, such as natural gas versus nuclear.

> Cost of new capacity is simply total cost to construct a generator divided by the capacity ($/kW).

To illustrate how this might work, I will compare the cost to construct a new nuclear unit with the cost to construct a combined-cycle gas unit. Using a very rough ballpark figure of $8 billion to build a 1500 Mw nuclear unit, one would arrive at a cost of $5,333 per kilowatt. This compares to a new combined-cycle natural gas unit of 1300 Mw at a cost to build of $1.1 billion, for a cost for capacity of $846 per kilowatt. So one can see pretty quickly that there is more to the analysis than the cost to build, or else no one would ever consider nuclear power.

> The levelized cost is an estimate of the total lifetime cost of the unit divided by the total lifetime output.

When a utility planner considers a new power plant, he or she analyzes all the costs for the lifetime of the plant compared to the expected output. Comparisons for what type of plant to build are based at least in part on the total costs divided by the total output, which is called a *levelized cost*. The fuel source may have already been chosen on the basis of a need for diversity, or the analysis may be used to compare several optional fuel sources. There are other factors, of course, such as a guess (sometimes called a *forecast*) as to future fuel costs, financing costs, and a myriad of political issues. Eventually the analysis would boil down to a cost per megawatt-hour delivered. This would be the levelized-cost analysis.

Once the plant is built, the money that is already spent is called *sunk cost*. The analysis of what to do with an older plant would look at all the money that would need to be spent to keep it in operation, or a "going-forward cost." Going-forward costs would include any modifications for new environmental requirements; all the maintenance costs; and all the operating costs to keep the unit available. This cost can be thought of as the going-forward cost of capacity.

> The going-forward cost is an estimate of all the costs to keep a unit operational from the present to the end of life, divided by the expected capacity ($/kW).

Another important economic term, and one often used in close to real-time operation, is the sum of all relevant costs directly related to the production of electrical energy. This factor, which is called the *incremental cost*, consists only of variable costs related to the electricity-production process. Incremental costs would include the cost of fuel burned, the

> Incremental cost is the sum of all costs directly related to production of energy ($/Mwh).

assigned cost (an estimate) of wear and tear on equipment, and the specific maintenance that can be variable due to production. It might also include a factor for fees for environmental discharges, and taxes that are assessed specifically for output. Keep in mind that the incremental cost for a particular unit can vary depending on the level of output. Incremental cost is expressed in dollars per megawatt-hour.

Technical Terms and Concepts

I want to define a few of the technical concepts for generation that are important to operating the grid. In this section I will address the following:

- capacity
- start-up time
- ramp rate
- minimum output
- reactive capability
- stability
- black-start capability

> Capacity is the maximum amount of power that the unit can produce, measured in megawatts.

The most important technical concept is the capacity available from the unit. Capacity is the maximum amount of power that the unit can produce, measured in megawatts. The capacity rating for a fossil fuel or nuclear unit would be based on continuous availability. However, capacity for wind or solar generation would be based on the maximum amount of power for certain conditions. So the total available capacity for a system or a fleet of generators could be a variable estimate.

> Start-up time is the time needed for a generator to get up to speed and connect to the grid.

Another concept that is important for generators is their time to start up, that is, the time it takes them to get to a speed so they can be connected to the grid. This can be anywhere from minutes (for some hydro units and gas turbines) to hours (for large fossil fuel [coal or oil] units). Dominion Virginia Power had a large oil unit that ran so infrequently that the only way to start it up was to call in crews to work overtime. That unit was listed as requiring three days in order to start up.

Related to this start-up concept is the time it would take to ramp the power output of a unit from a low rating to full rating. This "ramp rate" is expressed in megawatts per minute and is very important when considering how best to use available units to follow a varying load (called *load following*). The concept of having a fleet of generators that can be controlled so that the varying load is met minute by minute is extremely important to operators and to the grid. In fact, following load is a requirement in the standards.

Another factor for many generating units is the minimum output. This is a power output below which the unit is too unstable to continue to operate. For example, when a large coal or nuclear unit is started, it gains speed until it is connected to the grid (the speed it needs to get to is based on the type of generator—but the electrical output is 60 Hz). Once connected to the grid, the unit immediately begins ramping up its power output until it gets to the minimum power level. To put some perspective on this, the minimum output of a 500 Mw coal unit might be 280 Mw. Once the unit is at the minimum power level, it is ready to proceed to full load or to be controlled in a load-following mode. The minimum power-output level is important to keep in mind when evaluating a generation mix. At low load periods such as overnight, many of the coal units might be operating at minimum load. I'll give examples of the minimum power-output concept in the "Nuclear Generation" section and in the final section, about generation mix, both below.

> Minimum output is a limitation for many of the larger-sized generators on the grid.

I've mentioned the concept of reactive power in chapter 2 and how it is important to have reactive capability near load centers to control the system voltage. Reactive capability of a generation source is an important factor to consider. Also, there may be a trade-off between reactive power and real power, such that the unit's power capacity may be limited when reactive output is required. Since events happen very rapidly on the grid, a large portion of the reactive capability of generators must be on some form of automatic control. The analyses that transmission operators do with respect to the single-contingency rule are based on an assumed automatic action to increase or decrease reactive output quickly. If the control were done manually, the reactive assistance would be too late in many cases.

> Reactive output capability for generators is crucial to the reliability of the grid. Also important is the capability of automatically controlling this output.

In chapter 2 I brought up the issue of stability. Recall that the issue is the ability of a unit to stay in lockstep with the grid during certain electrical (and mechanical) perturbations. For example, the unit should be able to stay in synchronism if there is an electrical fault a certain distance away on the grid, as long as the fault is cleared in a reasonable amount of time (measured in a few cycles on a sixty-cycle-per-second system). When a nearby fault occurs, in many cases the unit will speed up, as the fault has taken away the resistance that the unit is pushing against (this is like having someone suddenly open a door from the other side as you are pushing the door). The distance away, the type of fault, and the amount of time to clear are variable criteria that the grid owners would specify. Control equipment can be added to the unit, or to a combination of units, to aid in the unit's ability to maintain stability.

> Stability of a generator is the ability to remain in synchronism with the grid during disturbances.

The system operator would like for the fault to be cleared and the unit still to be online and putting out power. Most of the bigger units are able to withstand these nearby faults and remain synchronous as long as the fault is removed in a reasonable amount of time. If not, then the units should have some form of protection to trip them off rather than subjecting them to the grid if they are not synchronous with it. Smaller units with less mass and less momentum may have a problem staying in synch and may have to be tripped off.

One more variable that may apply to a small subset of generators is the capability to restart a dead system. Many of the units that we will look at below require off-site power in order to start up and get to the speed needed to connect to the grid. In fact, some of them require a lot of off-site power, such as a nuclear unit, which may need 40 megawatts or more in order to start. But if the grid is totally blacked out, there needs to be some generators capable of producing power without off-site power. These units are typically small, such as hydro or gas turbines. All transmission owners/operators on the grid are required to have black-start plans, and operators are required to have training for implementing these plans, preferably with system simulators.

How many units with black-start capability are needed? The amount of black-start capability, both in numbers of units and total megawatts, is a complex matter that requires careful consideration. A restart plan would have to consider the amount of off-site power needed by the first tier of units to start after the black-start units have started. A valid plan would then work through contingencies of pathways to get the power to each of these units. If one also assumes damage to equipment during the blackout event, one sees that this whole exercise becomes tedious. Also, as companies retire their old, inefficient generators, they are also retiring units that are black-start capable. This is a concern that the planners must keep in mind.

> Black-start capability is the ability of a generating unit to start up and generate power without a need for off-site power sources.

Nuclear Generation

I will start with nuclear generation as a source of energy for the grid. In this section I will discuss the following significant characteristics of nuclear generation:

- typically large multiunit stations
- expensive to build, but very low incremental cost
- regulated by the Nuclear Regulatory Commission
- nuclear fuel cycle and outage planning
- load following and how to avoid backing down a nuclear unit
- nuclear core safety and the need for off-site power
- long time to start up

As we know them today, nuclear units are fairly large (e.g., 900 megawatts of output), and often there are multiple units at a single site. These units were extremely expensive to build, and the infrastructure (technical staff, licensing, spent-fuel storage, etc.) needed to operate them is very large, yet the incremental costs of production are very low.

The Nuclear Regulatory Commission (NRC), which licenses the operation of commercial nuclear power plants in the United States, regulates these units and, unlike FERC, has the authority to issue orders without industry input. The original license issued was nominally for forty years. However, the NRC determined that an extension of twenty years to that initial license was practical and safe. Many nuclear units in the United States today have already received extended licenses.

About every eighteen months, a nuclear unit must be shut down for refueling. During these refueling outages, the unit must be cooled. Then a portion of the nuclear fuel is replaced (as my nuclear friends have explained to me, about one-third of the fuel is replaced with new fuel while the other two-thirds are rearranged in the core). Refueling can take thirty to forty days if the system operator is lucky, sometimes a lot more if problems are encountered. Refueling outages are expensive, and detailed, hour-by-hour schedules of the work to be done are typical. The outage dates are planned well in advance and, in the case of Dominion, do not correspond to peak load seasons (summer and winter). Larger utilities with more nuclear units occasionally have to schedule an outage during a peak season. During these times, the nuclear operators at Dominion would provide the transmission operators a five-year plan for the refueling outages for our nuclear units, which we counted on for planning system operations. These plans were treated as gospel, and everything had to come together to meet the refueling schedule.

Here's the problem: the energy available in the uranium fuel is a fixed amount that it would be wise to use before it has to be removed. The plan for the outage has taken into account the amount of energy remaining in the fuel so that most of the energy is used just as the refueling outage begins. But if the unit is off-line for an unplanned outage for several weeks due to an equipment malfunction, when the scheduled outage date arrives there is still that much available energy left in the fuel. The value of this energy, which will be discarded if the refueling starts on schedule, has to be traded off against the cost (millions of dollars) in a delay of the planned outage. If the outage was planned to end in, say, late May, the system operator's position would clearly be to refuel on schedule and then dispose of the energy not used. The system operator wants the unit online during the summer months, but a decision is not always made on that basis. If the unplanned outage is very long, the pressure to use up the fuel in the core is even greater. This is also the reason that the nuclear folks don't want their units to be put on load control. If the unit were on load control, then the energy left in the fuel would likely be discarded when the scheduled outage started.

So, nuclear units usually run at 100 percent power level when they are able. With a few exceptions, nuclear units are not used for load following, since this would be a noneconomic way to operate them. It isn't that they can't follow load; it's just that the economics of nuclear operation don't favor it. Another way to look at this is to consider that the incremental cost of nuclear power is probably the lowest of all available units except renewables. Also, it can be very difficult for a system operator to explain why a nuclear unit that incrementally costs, say, $10 per megawatt-hour was backed down to follow load when a fossil fuel unit that costs, say, $100 per megawatt-hour was still running. There is an explanation for this situation, which I will attempt to describe.

When I started working in the system operations business, one of the first things my friends in the nuclear group told me was to never, ever order a nuclear unit to back down power output (such as to follow load down in the middle of the night). I can say that this never happened on my watch, although there were instances when we came very close. This is what happens: After the system operator studies all the relevant inputs, such as expected load for each hour of the upcoming week, the available units and their start-up times, and minimum capacities of all the units online, a plan is derived that meets the load at the least incremental cost. An operator then checks the total generation online at the time of the lowest load in the middle of the night (nuclear units at 100 percent; renewables at whatever they can produce; coal units at something less than 100 percent; etc.) to be sure that there is load-following room in some of the generators. Remember that the fossil fuel units online must be within their minimum and maximum ratings. The study case looks okay for the upcoming week.

Then the load overnight is significantly less than forecasted. However, the forecasted peak the next day requires all the units to be online in order to be available for meeting the demand. After all the fossil fuel units are ramped down to their minimum output, with the nuclear units at 100 percent and the load still decreasing, the options become few (remember that balancing authorities are supposed to match load and generation). These options are as follows:

- Take a fossil fuel unit off-line. (This would present another problem, because the start-up time is too long for the unit to be back in time for the peak the next day.)
- Sell excess energy to a neighboring system. (Usually when the load reached a low level for Dominion Virginia Power, the same was true for all our neighbors. Our neighbors didn't want to buy our excess energy, as they had the same problems. Many times, energy was simply given away to any utility that would take it. At times, no one would take it.)
- Load-follow with a nuclear unit. (This is not a popular outcome for the people responsible for the nuclear fleet.)
- Ignore the issue and simply overgenerate for a few hours (essentially putting energy onto the grid with no buyers, thereby raising the grid frequency). (This is noncompliant with NERC standards, since BAs are not supposed to stay in this kind of mode. Of course many BAs did this, as a look at overnight system frequency charts would show. Remember the discussion about the time clock error in chapter 1? This is one way that the interconnection can easily build up some error—by running fast.)

This is when the operator gets into the situation mentioned previously, where he or she has to back down a nuclear unit's output while a coal unit is generating at its minimum output, but at a cost of $100 per Mwh. There might be an option to take the coal unit off-line and replace it the next day with a more expensive turbine (with a short start-up time) or with purchases from neighboring systems. All of these options add cost, though.

Nuclear units have a particular safety concern that other units do not have, namely, the safety issue of the nuclear core. When a nuclear unit is tripped off-line, the automatic features in the unit will stop the nuclear reaction. However, there is a great deal of heat left in the core, which needs to be cooled. There is a lot of equipment that is designed to provide core cooling, including on-site diesel generators that can power this equipment in the case of a loss of the grid connection (called a *loss of off-site power*). A loss of off-site power is a nuisance for any unit, but with nuclear units a loss of off-site power is extremely serious. After a loss of off-site power, the safety of the core is dependent on the last line of defense to provide the core cooling: the diesel generators.

So it is crucial to maintain the grid supply at a nuclear station at all times. In fact, the NRC has a rule that the unit must be shut down within a certain time frame if there is only one transmission line left tying the unit to the grid. This can lead to some extremely busy transmission people when there is only one line left in the substation feeding a nuclear unit. And this always seems to happen when the situation on the grid is most dire, when the grid operator is counting on the unit to keep the grid up, so it is a very stressful time for all.

Another characteristic of nuclear generation is that when units trip off-line, it takes a long time to get them back online. The problem that caused the trip must be clearly determined and fixed before the unit is cleared to start up. The NRC is likely to get involved in unusual events and may find a slew of problems that must be fixed. This is, of course, due to the safety concerns, as I mentioned above. My experience was that anytime a nuclear unit tripped, it was better not to plan to have it back for several days, regardless of the risk to the grid. By comparison, the fossil fuel units could return in a matter of hours (sometimes) after minor events.

But for the most part, as long as the nuclear units are humming along at 100 percent output, they are nice units for a system operator to have. When a unit runs seamlessly, it is called a *base-loaded unit*. This means that it runs at full available output all the time regardless of the load. In the United States there are about one hundred nuclear units, which produce 25 percent of the electric energy consumed (this information is from 2012; there have been announcements of nuclear unit shutdowns since then).

Coal Generation

Coal generation is virtually everywhere in the Lower 48 states. The output capability of coal units varies over a wide margin; many of the largest are in the range of 500–600 Mw capacity and bigger. In 2011, coal was used to produce about 45 percent of all electric energy consumed in the country. In this section I will discuss the following characteristics of coal generation:

- environmental restrictions resulting in shutdowns
- possible improvements to environmental factors
- coal units providing needed outputs for a reliable grid
- start-up considerations
- ability to store the fuel on-site
- tube leaks

The US Environmental Protection Agency (EPA) environmental restrictions issued since 2008, including the Clean Air Act, Section 112, Utility Air Toxics—Title I of the Clean Air Act—National Emission Standards for Hazardous Air Pollutants (NESHAP), will eventually result in the shutdown of a number of coal units. This is especially true for the older and smaller units, for which it is not practical to spend the money on the upgrades needed to meet the regulations. NERC predicts that as much as 78,000 Mw of the nation's coal, oil, and gas capacity could be shut down over a ten-year time period, according to their 2010 Special Reliability Scenario Assessment. Much of the lost capacity will be made up with combined-cycle natural gas and wind units. In 2012, coal was used to produce 37 percent of electric energy consumed in the United States, a steep decline from 2011.

As of this writing, there is no universally applied "cost" of carbon dioxide emissions. It has been considered, though, and a lot of work has been done to see the potential impact of such a "carbon tax." Depending on the "cost" of a ton of CO_2 emissions, the impact on a coal plant could be anywhere from less running time (since other units would be cheaper to run based on incremental cost) to the complete shutdown of the unit.

The typical coal unit today (not the ones that are going to be shuttered as per the discussion above) emits over two thousand pounds of CO_2 per megawatt-hour. This is approximately twice the emission of a new natural gas unit. Can coal be used "cleanly" to produce electrical energy? If some technology can significantly reduce the carbon emissions from burning coal, then the future of coal as a fuel is secure. There is enough coal in the United States to power our energy needs for years (I've seen estimates of three hundred years), if we can find a way to use the energy without also harming our environment. If not, the future may not be very bright for present-day coal units.

There is a great deal of research being done to develop a cleaner coal-burning operating unit. Removal of mercury, nitrous oxide, sulfur dioxide, and other particulates has helped the more modern plants, and as mentioned before, the older ones are being retired. Also being considered is a chemical cleaning of the coal prior to combustion. It is very hard to imagine a coal plant that has no emissions, however. The removal of the carbon dioxide emission is the concern that the proponents of climate change will demand. One approach being studied is called *carbon capture and storage* (CCS). The plan would be to capture most of the CO_2 in the discharge gas and then "sequester" it to some underground storage facility. Even if this works and the CO_2 stays captured underground, the cost of doing this may upset the economics of coal use in electricity production.

Coal units are excellent providers of reactive power. They are very capable of load-following operation, and they are quite capable of providing the frequency-dependent variations (boosting output for a slow frequency) that the system needs when frequency declines after the loss of a generator. The turbine generators on these units are typically massive, meaning that they have a great deal of angular momentum and can usually survive electric perturbations that often occur on the grid. It is for these reasons that replacing these coal units with wind units is worrisome for many system planners and operators.

When called upon to run, a coal unit goes through a start-up period of variable length, depending on the unit design and other things. These units typically take hours to get online, after which they need more time to get up to full power output. This results in the need to plan well in advance for the start of a unit. Another economic factor that applies especially to these units is the cost to start up. Starting up these units is nothing like starting a car. The process is to heat water to steam and then use the steam to turn a turbine, which turns the generator. It takes quite a bit of heat to convert a large body of water to steam. This is done by burning fuel, sometimes oil and coal. This start-up fuel does not produce any output to the grid, since the unit is not even connected to the grid until it is spinning. In addition to the fuel burned is the cost of running a plethora of other equipment that supports the process. So for these units, the start-up cost can be significant.

When taking into account the time to reach full load and the cost of starting up, a big question arises: whether or not these units can be cycled daily. That is, can the unit be shut down overnight when the output is not needed and then restarted the next day? If not, then the unit will likely be running at its minimum power level overnight.

One distinction about coal is the fact that there can be fuel storage on-site for extended runs if, say, the coal miners or the railroad workers go on strike. In contrast, gas turbines are dependent on the pipeline, as any on-site natural gas storage would not amount to much run time. The coal pile is carefully managed at all times, from the standpoint of having no less than a certain number of days of full load unit output available. If the coal pile ever gets below some minimum level—the result of delivery problems, for example—the system operator could change the dispatch order of units to conserve coal. There is also a maximum storage limit, which is set by physical limitations and by financial limitations, since money invested in coal storage is not making a return.

I wasn't around the control center for long before I heard the dreaded phrase "tube leak." A tube leak on a coal plant is the nemesis of a coal plant's existence. Since heated and pressurized water must flow through the boiler (where the fire is) in most of these plants, eventually the tube bursts, creating a leak of superheated water and steam and leading eventually to the unit's being taken off-line. How many times on a hot summer afternoon have I heard these two dreaded words? On a hot afternoon, the question the system operator asks the unit operator is,

"How long can you keep it online?" If the leak was small and stayed small, perhaps the unit operators could nurse it through the afternoon peak. Unfortunately this was not often the case. The good news here is that the outage does not usually last for more than a few days.

Natural Gas Generation

Natural gas generation has become the preferred source over the last decade or so. The units cost less to build than nuclear or coal units, and they can be completed in less than half the time (a combined-cycle plant takes eighteen months to build and install). In this section I will discuss the following characteristics of natural gas generation:

- trends in natural gas generation
- different types of gas units
- load following, start-up, and ability to cycle daily
- availability of gas

The largest percentage of new units built today, or committed to be built today, are combined-cycle gas units. The fuel cost is looking very cheap, especially now that the fracking process is getting at huge deposits of natural gas in shale formations. In 2010, natural gas accounted for 24 percent of the generation in the United States, whereas in 2012 natural gas accounted for 30 percent. This trend is continuing now into 2016.

In a natural gas process, the gas is burned and the hot gases drive the turbine causing the rotation, much like in a car. The older units were just this simple, and in fact we call them *simple-cycle units*. Much more efficient units have evolved, however. The units that are built today (for the vast majority of cases) are combined-cycle units that take the waste heat from the output of the first turbine and then use it to run another turbine. Actually the recovered heat is used, along with additional burned gas, to heat water to steam and to run the second turbine from the steam. Efficiency levels on these systems can be pretty amazing, especially if they run at full load.

But as discussed in previous chapters, every unit doesn't get to run at full load. Some units must follow load. The simple-cycle units are not good at following load, although they can start up quickly. The combined-cycle units are better at following load, but they have the usual restrictions on minimum load levels. The arrangement can be more complicated than one gas turbine and one steam turbine to make up a unit. Often there are at least two gas turbines for one steam turbine. Other arrangements are quite possible, and each of these arrangements will have minimum and maximum output points. Within those restrictions, the units can follow load.

In addition to a combined-cycle gas unit's ability to follow load is its ability to cycle on and off over a twenty-four-hour period. Not only can these units start up fairly quickly, but also they have low start-up costs. This gives flexibility that is needed by the folks who need to match generation to load at the system level. Simple-cycle gas turbine units are excellent at cycling, have very low start-up costs, and have a very low start-up time (sometimes just a few minutes). In return for this great flexibility, these gas turbines will typically have a minimum run time, which means that they can be online quickly for handling a peak load or for replacing a lost unit, but they demand a few hours of actual output.

From a system operator's perspective, the one nagging concern with natural gas generation is the availability and deliverability of the gas. There is not much storage (if any) of gas on-site for most of these units, so they are dependent on the pipeline having gas in it at the delivery point to the plant. If the local grid has 5–10 percent of energy sourced from natural gas and the pipeline is lost, a system operator can probably make up the difference. After all, that is what reserves are for. But consider the risk of having 30–40 percent of generation from natural gas. Hopefully, this generation will be scattered around several different pipelines. Another concern that is expressed from time to time is the growing dependence that a lot of utilities have on the future of natural gas. That dependence is not only on the availability of the commodity, but also on the continued low cost.

Oil Generation

Oil generation is not economical in today's energy environment, so there won't be much new construction in this area. The amount of oil generation in 2010 in the United States was less than 1 percent (in 2000, by comparison, the percentage of oil generation was nearly 3 percent). Oil units are similar to coal units in that they have boilers, tubes, and tube leaks. Some smaller turbines burn oil, or some refined derivative, and have similar characteristics as small gas turbines (quick start, but not efficient). Large oil units, as the ones I am used to, burn crude oil, which is used to make steam. The rest of the process is just like that of a large coal unit.

Hydro Generation

Hydro generation is great to have available, and system operators love it. In this section I will discuss the following characteristics of hydro generation:

- great support for grid operations
- fuel limitations
- flood control and drought concerns taking priority

Hydro units can get online very quickly (within minutes), can provide reactive output, are able to follow load, and are very reliable. In 2010 they provided a little more than 6 percent of the generation in the United States. The fuel is essentially free and renewable, and the lakes formed are great for recreation. Unfortunately there are few, if any, ideal locations left for a large hydro facility in the United States. Therefore, the new construction of hydropower facilities is very small, 2 percent of the total planned for 2013, for example.

The hydro units are very useful as generation reserves since they can start up and get to full load very quickly with virtually no start-up cost. If a large coal unit trips off-line, it is likely that hydro units will be used to rebalance the generation to the load. This would then lead to a situation wherein the hydro unit would need to be replaced with something else, since there would not be enough water to keep the hydro unit running very long.

This leads to some important points about hydro generation that are different from generation for other renewables. First is the fact that the fuel available to a hydro facility is limited. From the system operator's perspective, this is the biggest drawback of a hydro facility. Unlike coal- or gas-fired units, hydro units are dependent on having water in the reservoir above the unit. The typical hydro plant was designed such that the units are used for peaking power. Put another way, the amount of water available to run the units generally limits operation to a few hours of run time a day. This mode of operation takes advantage of the fact that generation capacity and energy are more valuable at peak than at other times of the day or night.

Another important issue with hydro is the fact that the lake, the generators, and the dam are quite often part of a larger project for flood control. When hurricanes or droughts are in the forecast, the management of the water becomes first priority. This fact can have an impact from time to time on the operation of the plant. Specifically, when snow melts in the spring, or when the late summer or early fall brings hurricanes, the water stored is reduced to make room for huge inflows. Sometimes this water is "spilled," meaning that the water is released without going through a turbine. Operators hate it when this happens.

Wind Generation

Wind generation has become very popular over the past decade. In this section I will discuss the following characteristics of wind generation:

- success of wind generation in spite of economics
- typical installations
- load following using a free fuel
- variability of wind and the value of wind capacity
- support for a reliable grid: reactive, frequency response
- maximum amount of wind generation on the grid

The total wind energy output in the United States in 2013 was 4.13 percent of the total energy used, according to EIA. Compare this to the 1.5 percent that wind produced in 2000 and one can see the growth. The total wind capacity installed in the United States as of the end of 2012 was 60,007 Mw, also according to EIA. Wind generation has the following advantages: it is completely renewable, the fuel is free, there are no emissions, it is encouraged by our state and federal government, and—most important of all—it is subsidized. A lot of people don't realize the fact that today wind generation is not economical compared to gas, coal, or nuclear generation without the government subsidies. If carbon emissions are taxed, then the economics can change, but this has yet to happen in the United States.

I will explore some of the technical characteristics of wind generation from the perspective of the system operator. Wind generation at the utility level (i.e., large) is made up of "wind farms," which consists of many individual units. These units may be around 1 to 6 Mw each, but the important thing is that they generate electricity at a lower voltage than the grid uses. The outputs of multiple units are connected together and go through a transformer to raise the voltage to a level that can be tied into the grid. A wind farm might have fifty, one hundred, or more individual units. The total output is what the system operator sees, as the output of each individual windmill is not important to the operator.

Since wind is a free fuel, does it make sense to follow load with wind energy? In order to follow an increasing load, the units would have to be operating at less than their capability so that they would have room to increase their output. And similarly, in order to follow the load down, wind units would have to reduce output below the total capability. Controlling wind in a mode to follow load is losing free energy, since the wind can't be stored. Instead, wind units are typically operated at the maximum output that the prevailing winds will allow.

But I want to point out that the wind machine's output can be controlled if necessary. And if conditions require, or if the economics dictate, the units can be controlled to follow load. The economics in this situation would be the value in a transparent market of the ability of a generator to follow load. This value may be discovered by bidding to an independent system operator who requests bids for some amount of load-following capability. But if wind is used to follow load, then potential renewable energy credits are lost, along with the wind energy that is not used. What will decide how this goes is determining the costs to benefits, that is, the value of the energy and the renewable energy credits versus the value of following load.

The biggest operational issues that come with wind generation are the variability and the unpredictable nature of wind. Many studies have been undertaken to understand the daily and seasonal variation of the wind in different parts of the country. If that variability is generally known, the system operators can learn to work around it. Where the problem comes in is with the tendency of the wind to be hard to predict on a micro scale. More specifically, how much output can a particular wind farm produce over the next ten minutes? It is quite possible that operators will find it necessary to run several fossil-fuel-fired units at a low level, ready to step them up if the wind fails. This operation of fossil fuel units at low levels costs extra. In addition, there would need to be other units capable of backing down output if the wind blows more than predicted.

There is one important point that I should make about the differences between stated capacities of these wind turbines and the actual delivered energy at any specific time. It should be obvious that the units can only produce energy to the extent that the wind is blowing. Typically they are sized for a high-wind day, which probably doesn't happen often—and almost certainly not on the hottest peak day of the year. This fact results in the application of a capacity credit for the wind farm that is considerably less than the total capability of the individual machines. For example, at one time in the PJM area, wind capacity was discounted to 14 percent of the total installed capacity of the farm (8.7 percent in Texas). If a company paid for the construction of 100 Mw of wind turbines at a farm, plus all the other equipment that goes along with it, that company would only get a capacity credit for 14 Mw (or 8.7 Mw in Texas). This significantly lowers the economic value of these generators.

There are a few other technical parameters that I'll address. First of all, and as discussed in detail in chapter 2, the wind turbine machine can produce reactive capability to help maintain voltage. In fact, the farm can be enhanced with electrical devices at the substation that can even boost this capability if necessary. In the case of controlling the output of the unit, especially in the decrease-power direction, the unit's response would be extremely fast. In the case of load following in the increase-power direction, if wasting some energy is acceptable, these machines can be very responsive. So the start-up time, ramp rate, and minimum load issues are no concern for wind generation.

The ability of the machines (or the farm in total) to respond to grid frequency excursions depends on the design, but the more modern machines now being built are capable of meeting this requirement—if the loss of some energy is acceptable. Practically speaking, however, spilling energy all the time would be the only way for the machine to be available to increase output for a decrease in system frequency. I doubt that this is going to happen. As more wind generation is added to the grid, the immediate response of the grid to low frequencies will decline.

Stability for nearby faults in the grid might be a concern for wind farms in some severe scenarios. This is because the units are small and light and have a tendency to speed up under severe conditions, possibly losing synchronism. However, the units can be automatically controlled to quickly regain synchronism and keep on producing. As far as I know, there have not been any major issues with wind farms with respect to stability for nearby faults. Automatic controls (power system stabilizers) can also help to avoid situations where the units try to follow perturbations in the system.

Since the inception of significant wind generation, it has been the practice to run the wind farms at the maximum output based on available wind. Many of the operators I know have expressed a concern about how much wind energy the grid can accommodate. Think of it this way: if we reach a goal of 20 percent of our energy produced from renewable resources, then a large number of fossil-fuel-fired units will be taken off-line. Most of the units that are replaced are the very units that we depend on to follow load today. Why? Because it makes sense economically to follow load with the more expensive units while base loading (running at maximum available) the cheapest units, such as nuclear. Therefore, as wind energy grows to significant levels, older, high-incremental cost units will be retired—or they will need subsidies to continue to operate. There will be times, as the amount of wind generation continues to grow, when some wind generation will have to be used to follow load. The need to back down wind generation can be minimized with the use of some smart-grid applications or with additional ways to store energy.

What we have found is that there are several grids in Europe with 20 percent wind energy or more. It turns out that the wind is more predictable than we thought, especially if wind farms are scattered around a very large area (such as the area within a grid). The large area tends to smooth out the effects of local wind events such that the overall wind output is usually close to predicted. Also, the daily variations of the wind can be accommodated by adding other available units that respond quickly, for example, gas or coal, such that there are no major grid problems. Getting to this point in the operation of the system requires some new applications and much training, some of it on the job. Interestingly, though, data from Europe indicates that there is not a huge added cost burden on other units that are not being operated efficiently on account of the wind generation.

Solar Generation

Solar generation is also making some big inroads into the overall electricity-generation picture in the United States. In this section I will discuss the following things related to solar generation:

- types of solar generation and some statistics
- support for grid reliability

There are basically two kinds of electric generation with solar energy, a renewable resource. First is the use of the heat delivered by the sun to heat a working fluid (such as water) to steam in order to drive a turbine. This is called *concentrating solar power* (CSP), wherein fields of mirrors reflect the sun's rays to a central point where the fluid is heated. The second type of generation is where the sun's rays are converted directly to electricity, called photovoltaic (PV) generation. Unfortunately, both methods are more expensive than combined-cycle gas (CSP is more than twice the cost) or coal, on a levelized cost basis for the life of a plant, according to the Department of Energy (EIA) report *Annual Energy Outlook 2015*.

According to the Solar Energy Industries Association (SEIA), by the end of the third quarter of 2013, there were 10,250 Mw of "cumulative solar electric capacity operating in the U.S."[1] Also according to SEIA's website, there are thousands of megawatts of CSP in the works, especially in the Southwest. There are incentives that help the economics work out, such as an investment tax credit that is in effect through 2016. If a tax on carbon emissions were to become law in the United States, the whole economic picture would change. Also, the cost of solar PV units is steadily decreasing, which will change the economics and, potentially, the generation mix.

Solar generation has some operational negatives in addition to its high levelized cost. Available capacity in the best of cases would only be during the day, of course, and the units would not have much to offer for following load unless storage methods can be improved. There is not much chance that this form of generation is going to help with frequency recovery on the grid. Solar power counts as renewable energy. With the subsidies in place, there will be many new installations. This is where the application of new technologies might change things, if PV can be made more economical. At this point, according to the EIA report mentioned above, PV costs significantly less (on a lifetime levelized cost basis) than CSP.

Geothermal Generation

Geothermal generation is a renewable energy that is certainly sustainable—and the fuel cost is free. To gain some perspective on this type of generation, think of it as using some kind of working fluid to capture the heat from the earth and using that heat to run a turbine/generator. There are multiple ways of doing this. I don't have any experience with it and don't want to get into the details. According to the Geothermal Energy Association, as of 2012 the United States was the biggest-producing country, with 3,386 Mw of capacity in service. The levelized lifetime cost of geothermal units is high, however, with natural gas and coal coming in lower. But generally there are incentives that help make the economics work out.

---

[1] "Solar Industry Data," *Solar Energy Industries Association*, accessed January 3, 2016, http://www.seia.org/research-resources/solar-industry-data.

The generation itself would have many of the same operational features of coal units. Technically the geothermal units could be designed to follow load, react to frequency excursions, and be electrically stable. Whether they are operated in load-following mode is a similar question to the one examined during the discussion of wind generation. Since the fuel is basically free, does it make sense to back down generation? I suspect not—which puts this form of generation in the category of other renewables in that it is online at 100 percent of its capability (base-loaded). As I discussed in the section on wind, the more of this kind of generation that is online, the more challenging following load can be. Geothermal generation has posted some impressive numbers for availability, which would be a good thing for the system operator.

I was surprised to learn that there are emissions from geothermal plants. Existing plants' CO_2 emissions are one-eighth of the output of a conventional coal unit, according to the International Geothermal Association. However, these emissions can be tightly controlled in more modern units. There are also some dangerous chemicals in the heated water from geothermal sources (such as mercury and arsenic) that have to be dealt with (generally returned to the earth). And unfortunately there was at least one plant (in Switzerland) that was shut down when earthquake activity increased significantly within days of its start-up.

There is no question that geothermal energy holds a huge potential for future renewable resources in certain areas. At present, though, it still needs some governmental incentives if it is to compete. Just as I've said many times before, if the United States ever implements a tax on carbon emissions, all the economics will change.

Fuel Cells

Fuel cells, which convert chemical energy directly to electricity, have actually been around a long time, having been invented in 1838. With a continuous supply of natural gas and air, these things can supply highly reliable electricity. I don't see them being applied at a significant level in utility-scale generation, but they are ideal for small-scale generation. Residential applications could scale up significantly, although fuel cells have not made big inroads yet. With properly installed fuel cells, homeowners could effectively reduce their use of electricity during the day and sell any excess energy to the utility at night. However, the impact to the reliability of the grid could be negative if fuel cells make a huge penetration into the residential market. This is because there are concerns about their lack of reactive supply, their frequency support, and their stability, to name a few.

Energy Storage

When the market for wholesale energy first opened up in the mid-1990s, there were a lot of new marketers getting involved with the grid. As I discussed in earlier chapters, the rules in those days were not well defined, but the physics would not change for the sake of the new players. One hot day in our control center, the call came in from a marketer who had just purchased some energy but couldn't find a buyer yet, so he wondered if we could store it for him on the grid for an hour. Unfortunately, the grid can't store energy like a gas pipeline can store gas. We had to tell him no. He may have bought some energy that he donated to the grid.

New technologies for energy storage at grid scale could be a game changer for our energy market as well as for the use of renewable energy throughout the world. Storage has been touted by many as the key to the future of many types of energy supplies, including solar and wind, so I want to discuss some technical aspects of storage.

In order for energy storage to make a dent in the energy required at the grid level, the storage capability would have to be big. A system that peaks at 20,000 Mw can use up to 20,000 Mwh of energy in one hour. So storage systems that can store, say, 1 Mwh of energy may have some great applications for specific loads, but unless there are hundreds or even thousands of these systems available for the system operator to control, energy storage is not practical at the grid level. Battery systems are simply too small individually. This is not to say that small storage systems don't have their place, but it is to say that to find a realistic application at the grid level, there would have to be a large number of them used simultaneously. And such applications are potentially in the works, as I will discuss below.

There are two types of energy storage in service today that can store enough energy to be practical at the grid level. These are pumped storage (water) and compressed air. Pumped storage is a hydro facility that has a lower reservoir (below the turbines) and an upper reservoir (above the turbines). The idea is to pump water from the lower reservoir to the upper reservoir when the load is low and to run it back down through a turbine generator when the load is high. The compressed air idea is similar; there would be an underground vault of some kind that air could be pumped into during low loads, and then run back through a turbine generator when the load is high. In both cases the total storage capability would be measured in thousands of megawatt-hours.

Another form of storage that is worth some consideration is a large-scale application of distributed battery storage. The application comes up in the smart-grid discussions in connection with the infrastructure needed to support millions of battery-operated automobiles. Think of millions of batteries in cars across the nation—millions of cars each plugged into a controllable device for recharging at any given time. The batteries in all those cars can discharge through the recharger as well as be charged. This would potentially give us the scale for battery storage that I was talking about earlier.

I don't want to ignore flywheel storage of energy, either. As seen with batteries, flywheel storage may be very practical for some smaller applications, but it does not have the scale for a grid-reliability application. For example, it may make great economic sense for a wind farm to smooth out its output by using flywheels to store energy when the wind is blowing and the load is low. But the total energy stored is measured in a few megawatt-hours. This just doesn't do much for the grid itself. But, as in the car application above, if there were lots of these flywheels that could be controlled centrally, the method could be useful. I know I'm hedging a bit here, but I hate to say that flywheel storage will never happen.

What Is Green Energy?

The US Environmental Protection Agency defines green energy as energy that can be generated with no greenhouse-gas emissions from burning fossil fuels and that is not detrimental to the environment.

Green energy is a more restrictive term than *renewable* since renewables can include forms of generation that do cause some impact on the environment (such as large hydro facilities). Green energy generation includes wind, solar, geothermal, eligible biomass, and low-impact small hydro plants. Renewables would include these and also large hydro projects, that is, wave and tide generation. The difference is that renewable generators can have an impact on the environment, but the source of the energy is not depleted in the process.

If someone were interested, he or she could buy renewable energy certificates (RECs) that support the building and operation of renewable units. One REC is good for claiming support for the generation of 1 Mwh of renewable energy. However, it would not be the energy piece that is owned. The output of the renewable generator would be split into two pieces: one being the actual energy that went on the grid and comingled with the output of every other generator, and the other being the right to show support for renewable energy. Of course there is no way that the electrons from that renewable generator are going to find any specific load. But if someone wants to buy enough RECs to cover all their electricity usage so they can say, "All my electricity usage is renewable," I won't be the one to point out the law of physics.

What Is the Best Mix of Generation?

So after looking at all these options for generation, we ask what the best mix is for a reliable, low-cost, centrally controlled balancing area. This question is really answered in two completely different time frames: long-term planning and real-time operations. The best mix for the system over the next ten to fifteen years is different from the best mix for an operator planning to serve load through the next day or week. Following are the issues that I will be discussing in this section:

- the criteria used for the mix: reliability and economic feasibility
- using computer programs to analyze costs for multiple scenarios
- integration with independent power producers
- the basic types of generation: base, intermediate, and peaking
- seasonal load shapes that must be met
- parameters of the long-term study of generation mix

- operations analysis of short-term conditions
- example of a generation mix for a peak summer day

Perhaps before delving any deeper into answering the question of the best mix of generation, I should better define the criteria for the mix. Primarily, the overall goal is a reliable mix of generation. This would mean, in addition to maintaining a percentage of generator reserves above the peak load, a mix that is responsive to load so that the operating standards can be met; is dispersed around the system to provide flexibility to meet the load if transmission lines are lost; meets all regulatory requirements; and provides diversity in fuel type so that the grid is not dependent on one fuel type. Secondary is the criterion that the generation mix operates at a reasonable cost, both in capital investment and incremental costs. Regulatory requirements include things such as emission limitations, for some states a stated reserve margin, and for many states a renewable energy target. Granted, then, we are not looking for the least-cost operation but for a reasonable cost for a reliable system that meets requirements.

I will start with the long-term view. What is needed is a computer program that can take all the variables as inputs and provide estimated costs for operating the system. Of course, there are such analysis tools that can arrive at a total estimated system cost for a given set of conditions. The input conditions would include things like weather, fuel cost, load growth, and unit outage rates. These studies are then run for a range of inputs to see how these conditions will affect the overall system cost. But this would not result in one lowest-cost answer for all conditions. Instead, it perhaps would show that for certain probable ranges of inputs, there is one mix of generation that tends to outperform the others.

The advent of third-party independent power producers has complicated the long-term planning process. I say this because the independent power producers can decide to build a new generator and connect to the grid in the same area where the planning I'm talking about is taking place. The planners, having to include the plans of the independents, have figured out ways to take this into account. It simply takes away some control, leaving a lot of additional uncertainty in future plans. One of the biggest uncertainties arises when an independent power producer announces plans to build a plant and then later cancels the plan.

The units that are chosen to be in the pool of available generation are divided into three groups based on the incremental cost of production (not the levelized lifetime cost). These groups are base-load, intermediate, and peaking.

The base-load units are those that are so low in incremental cost that they run at 100 percent of their capability around the clock. Nuclear, wind, solar, mine-mouth coal, and geothermal generation sources are typically in this category. How much of the total load can be met by base-load units? The answer is a bit more complex than a simple percentage of the total. Here we get into the issue of real-time operations versus a long-term plan. That is, the long-term plan may have shown benefits for a large block of base-load units, but on a mild day in late spring the plan just isn't going to work. Why not? Because there will be more generation online than the load, and since all the generation is base-load, there is nothing that can be used to follow the load up and down. Selling the excess energy to another area is tough to do in the shoulder months (i.e., spring and fall), since everybody is facing similar concerns. A good mix would be somewhere around 10–20 percent of the generation online capable of load following during the lowest load of the night.

The intermediate units would be the units that have an incremental cost somewhat higher than the base-load units. These would consist of natural gas combined cycle, coal, biomass (for example, wood chip burning), and energy storage facilities. These are the units that will give the operator the flexibility to get through the seasons (peaks and valleys) without breaking the operating rules. Some of these units are capable of cycling during a twenty-four-hour period, meaning that they can be started and run during the daytime hours and then removed at night. However, there are others that will have a problem doing that kind of cycling day after day. These units will be set to run at their minimum load during the night. In the discussion above, I mentioned the need for having around 10–20 percent of the generation online for load following. This would come from the intermediate units, but the problem here is that the units that are placed on minimum overnight are no longer available for load following.

The last category of units, peaking, would be the highest-cost units (that is, the highest incremental cost). This category includes the simple-cycle turbines and other, very expensive units. These units would not run very much, probably just at peak load periods and during emergency conditions. (Hydro units with reservoirs are in this category since they are planned to run only across the peaks each day, even though their incremental cost is very low.)

The generation mix has to meet the load shape for the area in question, and it must do so effectively every day of the year. Diagram 5-1, below, shows a load shape for a peak summer day in a typical large area of about 20,000 Mw peak load. Diagram 5-2, below, shows a load shape for a typical peak winter day. While I'm at it, I also point to diagram 5-3, which is a load shape for the shoulder months. Notice the peak loads of these three very different days, but also notice the minimum loads on these days. An examination of the winter load's shape curve will show a good example of the benefit of a peaking unit. Since the peak load is such a short duration, maybe just one or two hours, it is far more economical to meet the top of that peak load with a peaking unit. Another important point to look at is the difference between the peak for summer or winter, and the minimum during the shoulder months. Keep in mind that our mix of generation must meet our criteria over the entire range of loads.

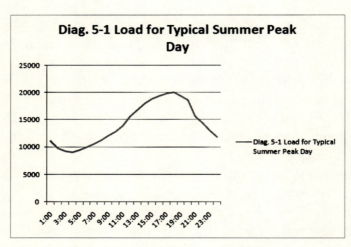

Load for a Typical Summer Peak Day

Living on the Grid

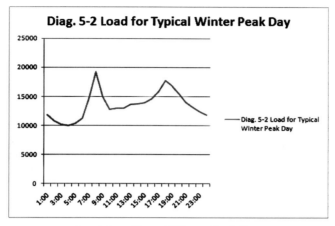

Load for a Typical Winter Peak Day

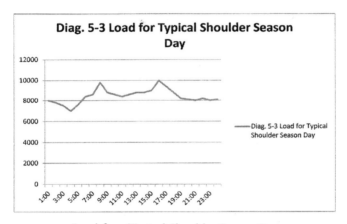

Load for a Typical Shoulder Season Day

I present this information to show some of the pros and cons of different mixes of generation. Base-load units typically have a higher cost to build but a much lower incremental cost. It would not be practical to build enough base-load units to meet the entire peak load since some of those units would only be needed 5–10 percent of the time. It is better to pay the lower cost to build a peaking unit and plan for it to run 5–10 percent of the time. It is also better to have units that can follow load that are not as expensive to build as the base-load units. These intermediate units may be off-line or at minimum output for significant periods of time, so their run time may only be 15–60 percent. The question is, are these intermediate units economically justified with such run times?

The answer to that question is complex. That is why we use a computer to run thousands of simulations to try to figure it out. The computer model would give us system costs to operate for various scenarios of load, unit outages, weather, and of course generation mix. The answer is a long-term assessment, as this is not just an annual cost evaluation. As we do these studies, we are not only getting an idea of the total system cost but are also getting a sense of the overall run times of each type of generation. This information could be used to determine which type of generation—base, intermediate, or peaking—is the best fit for the next investment.

I have described an analysis of the long-term view, which is not to be confused with the operational analysis. In the operating environment, there is a short-term view of alternatives, considering load shape, units not available, and all other constraints of the system and its equipment. Obviously, building a new plant is out of the question, so we only work with what we have. In this assessment, a least-cost plan would be made for which units run for each hour of each day over the study period. Start-up costs and incremental costs are used in this analysis, along with many constraints for operational concerns.

To better illustrate the short-term analysis I will show an example of planning a single day of operation. Using the curve for a summer peak day in diagram 5-1, I will try to fit a mix of generation into that demand. Diagram 5-4 shows one possible outcome. At the bottom of the curve are the base-load units, with nuclear as a fixed number and with wind and solar variable over the day. Next I have the intermediate units, which get close to maximum output during the peak, although there are some units available for load following at all times. Then I have the reservoir hydro units running for several hours across the peak and possibly a few simple-cycle gas turbine units for two hours at the peak.

But as a footnote to this plan I have developed, I also must be sure that the plan works for the minimum load overnight. The mix during the minimum load would include the base-load units at whatever they are capable of producing, some intermediate units at their minimum load, and a few intermediate units for load following. This would be an example of an operational plan, which in reality would have to include plans for the next few days as well.

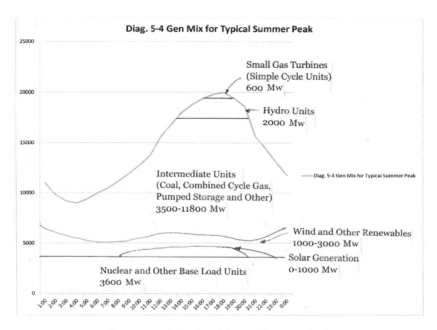

Generation Mix for a Typical Summer Peak

Cost-Based vs. Price-Based System: The Market

What I have described in this chapter is how the different types of generators might affect system operations. This has included a substantial discussion of cost issues. I have purposefully avoided the discussion of the wholesale market for generation. This is the way that I learned this business, from a cost basis first. In the real world, though, in a large part of the United States, generation fleets bid prices in a transparent market for energy. Price and cost are intricately connected, of course, but it is important to realize the difference.

How does a market for wholesale energy work? What is the impact on reliability? How does the market affect prices for electricity at the retail level? In the next chapter I will expand the discussion from a cost-based system into a price-based system.

Chapter 6

COMPETITION: THE MARKET

I once gave a tour of Dominion Virginia Power's System Operations Center for the president of Botswana, who was in Richmond for a meeting arranged by our governor (the first African American governor in Virginia) Doug Wilder. It turns out the man was interested in how interconnected systems operate. (His armed guards were not interested, however, as they promptly fell asleep.) He asked me how we priced emergency energy if a neighboring utility (and, in his case, a neighboring country) was in dire need. I told him that if they ask how much it costs, then it is not an emergency. Now, years later, I remember how he laughed about that. The truth is, though, that in those days the seller was limited to a price for emergency energy of 110 percent of the incremental cost to produce the energy.

Introduction to the Wholesale Market for Electricity

Is the wholesale market for electricity important to the understanding of how the grid works? My answer to this question is that the purchase and sale of electricity on the grid is an extremely important component of how the grid works. If the grid were only used as a reliability enhancement for otherwise self-sufficient companies, then a huge potential use of the infrastructure would be lost. The purchase and sale of electricity across the grid has an enormous value to the economy, although I've never seen any figures estimating it. At any moment, night or day, there are innumerable transactions taking place on the grid between entities. I think it is at least worth it to provide an introduction to how these transactions work. I will not go very deep into this topic, however, as my own experience is somewhat limited—and frankly, the electricity marketing business changes rapidly.

Before delving into the details of competition in the electric business, I will review operational criteria. Any competitive market that operates must have a framework that can be implemented while meeting the requirements for a reliable grid. The following goals must be met within defined parameters:

- match generation to load in defined areas
- maintain frequency of the grid close to sixty cycles per second
- limit current flows and voltages throughout the grid such that the loss of any single element does not cause collapse
- achieve an acceptable level of reliability for delivery to customers
- strive for an operation that achieves these goals at a reasonable cost (not necessarily the least cost)

The biggest cost to consumers in the electric business is the production of the electricity product itself. The "wires" business, both at the transmission level and at the distribution level, is a regulated business with monopolies providing the service at a regulated cost. As discussed before, there is little to be gained for customers when there are multiple companies with wires running down the street. There may be some places where meter-reading, billing, or even load-management programs are competitive, but savings to the customer for these types of competitive endeavors are relatively minor. So my focus in this chapter will be on competition in the production and wholesale trading of electricity.

Purchases and Sales of Wholesale Electricity the Old Way

I want to start the discussion of wholesale trading by presenting a high-level view of trading prior to the federal push for competition. By learning the old way of trading first, the changes that came about with competition should be easier to comprehend. In this section I will discuss the following:

- contracts used to transact purchases and sales
- economy energy

- wheeling of economy energy
- emergency energy
- capacity-backed purchases and sales
- unit power sales and minimum energy take or pay

In the old days, energy and capacity were sold on the grid by integrated electric utilities with bilateral contracts approved by FERC. These contracts were called bilateral because they were set up between two bordering utilities with tie-lines, and FERC was involved because any transfer of power on the grid was considered interstate commerce (except in Texas). Bilateral contracts governed day-to-day transactions as well as some longer-term deals that fell under the scope of the contract. Contracts that were more complicated would be negotiated and then filed with FERC for approval.

I'll start with the simplest example. In this example, there are two utilities that are interconnected with tie-lines. They have a bilateral agreement, approved by FERC, that allows for energy transactions between them. Utility A has a fleet of generators online with a maximum incremental cost of $30 per Mwh, with the ability to generate another 100 Mw before the incremental cost goes up. Utility B has reached a maximum incremental cost of $50 per Mwh and could back off the more expensive unit if they could bring in 100 Mw. Using the bilateral contract, the operators at utility B agree to buy 100 Mw from utility A for a cost of $40 per Mwh.

The type of transaction described above is for "economy energy." There is no capacity backing it up, and the energy flows only as long as buyer and seller are still satisfied with the deal. The expectation in this situation is that the utility buyer owns generation that can be used to make the energy but has realized a savings by buying from another utility. The pricing for this energy is very simple, as governed by the bilateral contracts. It is a split-the-savings arrangement. Both parties are happy.

Now imagine three utilities, tie-lines connecting utility A with utility B, and tie-lines connecting utility B with utility C. There are no tie-lines between utility A and utility C. If utility C wants to buy some energy and utility B doesn't have any available but utility A does, then C must ask B to do a buy-and-resell transaction where B uses its bilateral contract with A to buy the energy and then B uses its bilateral contract with C to resell it. But FERC approved these contracts to limit the amount that a utility such as B can make in this transaction. What utility B is actually doing is called "wheeling" the energy from A to C.

For years these economy-energy transactions were one of the main methods used by utilities to trade energy across their tie-lines. In fact, there was an expectation that the operators would implement the deals whenever possible. There was no need for management approvals; an operator could just do it. At any given time, there were economy transactions going on all over the grid. If any problems sprang up on the grid, it was easy just to cut the transactions. Of course, this assumes that everyone is playing by the rules, which means that buyers are only buying this product so as to allow them to back down higher-cost generation. That is, they have the capacity they need, just not at a low incremental cost.

This brings me to the next type of energy-only transaction that worked in the old days. If a system was generating all it could but load was exceeding generation, then that system was short on capacity. Assuming there is no capacity available to purchase, the next option is to buy "emergency energy." This product, similar to economy energy, was sold if available, but there were several differences in the terms of the transaction. FERC approved the bilateral tariffs' charge of "cost plus," usually a 10 percent adder to the seller's incremental cost, since there was no generation available in the buyer's system to split the savings. From an operational perspective, however, the main difference was that if the emergency energy was curtailed, then the buyer had to shed some load. Since nobody wanted that outcome, sellers would do all they could to maintain the transaction, including buying additional emergency energy from other neighbors and wheeling it to the buyer.

There were also capacity-backed transactions in our bilateral tariffs. In this case, the variable was the duration needed. Capacity for a day, a week, a month, or a year were all alternatives. We didn't have hourly capacity, as I recall, as it just wasn't practical to deal in such a short time interval. The amount of energy that went along with the capacity was usually up to the buyer. For example, a buyer may purchase one-day capacity but only take the energy over the peak hours of the day. In any hour, the energy could not exceed the capacity purchased. If the deal was for a one-day capacity purchase of one hundred megawatts, the buyer could take any amount up to one hundred megawatts in any hour.

This capacity was sold as a "slice of system," meaning that the capacity wasn't specific to any unit but was available from the total fleet of the seller. If the seller got into trouble on his own system, he had the right to curtail the transaction prior to his shedding customer load (of course, this may have left the buyer having to shed customer load). One interesting thing about these deals was the fact that the transmission needed to move the energy was bundled in the deal. This is not the way business is done now, as I will describe later.

In the bilateral contracts, FERC regulated the pricing of all these capacity transactions as well. The capacity charge was derived from a filing with the FERC of the all-in cost of the capacity. Energy was sold on a cost-plus basis similar to the emergency energy above. A seller could ask for less than the approved capacity rate but could not ask for more.

There were also more complicated, longer-term transactions, of course. These deals were negotiated between the parties, and then the contract had to be approved by FERC. In many cases the situation was that one utility would be building a new unit, larger than needed to meet its load growth for the next few years. For example, a utility decided that it made sense to build a 500 Mw coal plant even though it only needed 250 Mw for the next five years. The utility would like to sell the extra 250 Mw capacity (and associated energy) for the years when it was not needed on its own system. So the utility would strike a deal for a "unit power" sale of the 250 Mw capacity to another utility. This is not the same as a slice-of-system sale, as the unit sold must be online for this capacity to be available. One important consideration was that the seller would usually have no right to retain the unit (curtail the transaction) if the seller's system was in trouble. At Dominion Virginia Power, we were the buyer of one of these deals that went on for years. It included a requirement for a "minimum take" of energy. This means that the buyer was required to take energy up to a certain point at all times that the unit was available. Doing this ensured that the unit would stay online rather than be shut down for economy during low-load conditions. This was also called a "take or pay" transaction since the buyer was required to take energy (or at least pay for it) all the time.

This is as far as I'm going to go into the history of buying and selling energy in the United States. What is important to understand is the framework under which these deals were made and how that changed with wholesale competition. Transactions across the interconnections were for the purpose of enhancing reliability and improving economics to whatever extent possible, based on the perspective of the transmission owners. I dare say that no one was getting rich on these transactions.

The Difference between Cost and Price

Before the wholesale competitive market for generation was implemented, the incremental cost to run generation was used to determine a plan to operate a fleet of generators to achieve the lowest possible cost. This determination was typically performed with the use of a computer program. The program would spit out a seven-day least-cost dispatch of the generators available so that the load was matched every hour of the week ahead while meeting all reliability constraints identified. In those days, each generator's costs related to energy production were entered into the program. In real time, the generators were dispatched based on actual conditions at the time and on their incremental costs for energy production. The computer that controlled the actual real-time operation was different from the one that projected the week-ahead costs. However, the main point here is that both computers used detailed estimates of the incremental cost to run the generators.

After a wholesale market was implemented, the "costs" to run the units were no longer used to determine a least-cost operation. All generators were required to submit their "price" quote to the independent system operator (ISO). These prices could vary over the twenty-four-hour day, and they included a price for energy (in the form of $/Mwh). The ISO uses the price bid, along with constraints, to arrive at a dispatch plan. The objective is still the same: to achieve lowest overall "cost" while meeting reliability requirements. Instead of dispatching based on cost, now the ISO plans the hourly generation mix based on quoted prices from various generation owners. But costs are still important to the generators, who have their own strategies for bidding to recover their costs and to make a profit. For instance, a nuclear unit may bid a price of $0 for certain hours to ensure that the unit will run at full power for twenty-four hours per day. Since all generators that run are paid the same, namely, the rate of the highest winning bid, the nuclear unit would come out with a good return when such a strategy was used. There have been times, however, during low-load conditions when the market price turned out to be $0 for some periods.

The Products Traded at the Wholesale Level

The Open Access Transmission Tariff established a new framework for wholesale trading on the grid starting in 1996. The production, delivery, and sale of electricity now had to be unbundled and offered as separate products. This concept is important to gaining an understanding of trading at the wholesale level. In this section I am going to discuss the following unbundled products:

- energy
- capacity
- renewable energy credits
- reactive-power capability
- load-following capability

The most important product is the energy itself. Residential final users don't really care about the other "products," as it is just the energy that they want. Energy is what spins the meters. For residential customers, it is the basic commodity for which they are billed. The fact that all customers may want energy at the same time presents a problem to the producers, though.

The producers need to have the capability to produce that energy at the time it is demanded. That drives another very important product at the wholesale level: the capacity to produce energy. Capacity has value in the market, in addition to energy, in order to ensure that there is enough capability to produce energy when customers want it. If capacity were not valued and traded in the market, then there would be a lot more rotating blackouts going on.

Here is a simple analogy of this capacity concept. When someone rents a car, she or he is paying for the right to drive it, whether or not she or he actually does drive it. Most often the deal is to rent for a period of time with unlimited miles. In effect, the person has rented the *capacity* to drive one car out of the fleet of cars at the rental agency. But even if the car is not driven, the rental fee is due. This is very much like buying electrical generation capacity from an entity that owns a fleet of generators. If the owner had the right to take the car back at any time, then this arrangement would be equivalent to buying electric energy without having the capacity to back it up.

As mentioned in chapter 5, renewable generation can be separated into the energy, capacity, and another product with value called a renewable energy credit (REC). What gives value to the RECs are customers who are willing to pay more for their energy if it is generated by a renewable energy source. Where a state has set specific requirements for production using renewable energy, an integrated utility company may have to buy RECs to meet the goal, even if the renewable generator is in a different state and the energy is already sold.

Another product that has value is the capability to produce or absorb reactive power. As discussed in previous chapters, reactive power is a must to maintain system reliability and also must be available to ensure that the grid can survive any single contingency. Although reactive capability can be provided by equipment in a substation (such as capacitors), having the capability on a generator is preferred since it can be controlled quickly and more precisely over a range of operation. So, much like in the case of RECs, above, an integrated utility or an ISO may have to buy the reactive capability of a generator even though they will not be buying the energy.

Similarly, it is necessary for some of the units online to have the capability to follow load both up and down. Therefore, load-following capability has value and can be sold. In some markets the owners of generators that have the capability to follow load can bid offers to raise or lower output. The ISO then can decide on an amount of capability needed and may accept the lowest prices offered.

There are other products of electric production that can be traded in an open market, but rather than going into any more detail, the rest of this chapter is going to focus on energy and capacity. These are the most important products and are the ones most widely traded in all markets.

Changes Initiated by the Open Access Transmission Tariff

The main impacts of open access on wholesale trading include the ability to do deals over long distances, the unbundling of transmission and generation, and the removal of price restrictions for the parties doing the deal. After the tariff was instituted, no longer could a deal be made that included the transmission of the energy bundled along with the energy itself. In this section, I will delve a little deeper into open access and these changes. I will discuss the following:

- improved ability to trade over long distances
- market-based pricing versus cost-based pricing
- separation of affiliated transmission and generation functions
- securing a transmission path
- selling transmission paths
- firm transmission: FERC's interpretation
- a summary of the new framework

Although the bilateral contracts were maintained for some time (and some may still exist), after open access became the law deals to sell energy or capacity could be executed between two parties that were not interconnected. A transmission path had to be purchased separately, but once this path was obtained the transaction could go through several states and transmission owners.

During the time when this new framework was initiated, one of the marketers whom I knew did a transaction from the East Coast to the West Coast, just to show that it could be done. I'm pretty sure that the deal was not economically viable, but it proved that this new way of doing business worked. Before open access came about, the only thing a company could do was ask all its bilateral contract neighbors for energy and hope that if they didn't have it, they would ask their neighbors. This left us at Dominion Virginia Power unable, at times, to find the emergency energy we desperately needed—only to find out later that it was available on the grid (but too far away to find under the old, inefficient system).

Whereas the bilateral contracts that FERC approved were all cost-based, the new model assumed that a competitive market would set pricing for generation (energy and capacity)—and the tariff did not need to set prices for these products. Transmission pricing would continue to be cost-based, as would some of the other generation products I talked about above, such as reactive capability.

FERC has a test to assess whether a geographic area is competitive for wholesale generation. The test assesses market-share ownership of generation. Whenever FERC discovered areas where the market did not pass the test, the result of too much of the generation being owned by one party, they would not allow market-based pricing. This was especially exercised anytime a merger of two utilities was proposed. FERC had to approve these proposed mergers. To do so, they examined what was called *market power*. In many of the merger filings, FERC required the parties to sell a large part of the generation fleet. At the time this was developing, most of the big utilities were very interested in participating in this unregulated market, so any threat that they would not get market-based pricing was very serious. If they couldn't get market-based pricing, they would have to go back to a cost-based system similar to what I described earlier, and they would be limited on returns on investments.

After open access, the transmission owners had to treat all parties that wanted to use the grid equally. In fact, all aspects of the sale of transmission access were governed by the Open Access Transmission Tariff and, of course, by FERC. The folks who operated the transmission system had to be very careful not to give any preferential treatment to their affiliated generation folks, especially the marketers of those products. In cases where FERC found violations of this rule, the hammer came down hard!

The safest thing to do was to instill into the transmission operators the idea that they couldn't talk with the affiliate marketers.

The marketers who performed these deals were required to separately purchase a transmission path for the energy to actually flow. It is easiest to think of the transmission product as capacity, similar in many ways to generation capacity. A marketer would buy a path through a transmission owner for, say, 100 Mw for a month. The path would be used whenever market conditions made a profit and were simply not used otherwise. The transmission capacity was a sunk cost, but a marketer had to have it to make money. Many marketers discovered this when the energy market skyrocketed to several thousand dollars per megawatt-hour in the late 1990s. If they had not already reserved a path, they had to watch from the sidelines, because there was no more transmission available.

My experience during this time was on the transmission side. We at Dominion Virginia Power sold transmission to marketers based on our pricing and availability, which was posted on our OASIS system (see chapter 4). A lot of the paths we sold were through our system from the south to the north. Our tariffs had several durations of time for the transmission paths available, and although the maximum price was cost-based, we were allowed to offer discounts, but only if we offered the same discount to all market players simultaneously. But most important was that there were two levels of firmness that we sold. "Nonfirm" meant that the deal could be curtailed when constraints appeared. "Firm" meant that the deal flowed unless the grid was near to imminent collapse.

Here was the one of the biggest shocks to the old-time utility operators: FERC said that sales of firm transmission paths had to be treated with the same level of firmness as the retail load. That meant that if transmission constraints required operators to reduce the loading on the grid, then they had to apportion that reduction to retail load and firm transmission paths proportionately.

So imagine that a marketing company bought a firm path through the transmission owner's system. On a hot afternoon in July, the grid cannot stand all the flows. After all nonfirm transactions are cut, the operators determine that they need to cut another 100 Mw of flow. They could do that by entirely cutting the flow through the system that the marketer had arranged. But FERC said that such was not fair treatment; instead, the operators had to shed some load of their retail customers in combination with cutting a portion of the marketer's transaction!

Imagine how the state regulators in Virginia would have felt about a situation like this, where a marketer is buying power in South Carolina and selling it into New Jersey, and yet Virginia customers have to suffer a load shed while a major part of the transaction continues. Transmission operators found out quickly that selling firm transmission was risky. However, FERC required transmission owners to sell whatever they could and to post on the Internet their calculations to prove that the analysis of available transmission was reasonable (meaning that transmission capacity was not unduly withheld).

All of this major change in the way of doing business led to an entire new industry of trading and investing in unregulated generation. It spawned an improved way of providing energy needs in times of crises, and therefore reliability of the grid improved. However, it also led to the problems and challenges I have discussed in previous chapters. Some people even claimed that the new framework resulted directly in blackouts. My opinion is that the industry was moving too fast and didn't make the necessary changes in time to avoid some painful outcomes. Overall, now that several years' work in the industry has allowed us to develop the needed protections, I believe that the new market has led to better reliability.

In review, at a high level, transactions before open access were usually performed by two interconnected utilities using bilateral contracts with bundled transmission—fairly simply for the most part, especially when it came to energy. Prices were capped based on cost, and these transactions had a lower priority than serving retail load. In the new structure, deals are made by any two parties for generation capacity/energy at prices arrived at in an open market (they are not necessarily interconnected and do not necessarily own transmission). These parties buy transmission paths under the terms of the open-access tariff. If the transmission purchased is "firm," the transaction has the same firmness as retail load.

Markets Managed by Independent Organizations

As discussed previously, there are differences in how the markets are run throughout the United States. Many areas don't have a market as such, but they trade on the grid using the Open Access Transmission Tariff, as I have described above. In the area that I am most familiar with, the Mid-Atlantic, the market is run by an ISO. Generation within that area can sell into the wholesale market without the sellers having to pay a transmission fee to wheel the energy anywhere inside the footprint. There are still deals struck outside the footprint to bring energy to the market, and there are still transfers of energy from inside to outside. The rules in areas managed by ISOs are somewhat different from the rules in those areas without a market.

Generation or Transmission Built by Independents

The Open Access Transmission Tariff and the rules that go along with it have opened the grid to independent generators. Now a generator can be located where conditions allow and can buy a transmission path to a buyer virtually anywhere. Of course, the investment in generation is huge, and finding a long-term buyer makes it a lot easier to finance the investment. In any case, the structure of the operations on the grid would allow such a generator to function on a financial basis. Whether or not this can be done in an economically feasible way over the life of the plant remains to be seen.

Another potential change in this industry is independent transmission. As mentioned before, there are now independent transmission-only companies doing a viable business. Also available is the possibility of an independent company deciding to build transmission in a generally congested area normally thought of as being in an IOU's territory, and filing with FERC for approval of a tariff where the independent company can charge users for the alternative path. Of course, the independent company would have to take the downside of loop flows, just like all the transmission owners have to do. Whether this concept will ever come to full fruition remains to be seen. Getting approval to build transmission anywhere is so incredibly tough that it is hard for me to see this actually happening.

Chapter 7

GRID BLACKOUTS

Early in my career as a system operator, I was asked to help a young woman develop an educational video about the grid. Having been hired by an education company to make the video for elementary school kids, she wanted to interview me for a small piece about the importance of the grid to our everyday lives. She finally got around to asking me the question of what the impact would be if the grid blacked out. I gave my lengthy description of how different parts of our society would be impacted, and I included a technical discussion of the activities needed to recover the grid.

"Mr. Thompson," she said, somewhat apologetically, "this is going to be used for elementary kids, so you need to gear your answers more toward them."

I said, "Okay, let's do another take."

After the second take, she said, in a bit more disappointed tone, "Uh, that is better, but still probably too much detail and over their heads. Can we try it again?"

On the next try I said, "And if the grid is blacked out … that would be … uh … very bad."

I hope she found something from the interview that she could actually use.

Introduction to the Blackout Discussion

The northeast blackout of 2003 was the biggest electricity outage in North American history. There were over fifty million people without power, some for several days, in the United States and Canada. This event ultimately led to some important changes, which I have discussed, including the beginning of mandatory standards for the industry. Yet blackouts are the biggest thing that all the system operators on the grid spend their careers trying to prevent. How did this one happen?

The History of Blackouts

Before I get into the details of the largest blackout in North American history, I would like to put that event into some perspective. To do this, I will take a look at some of the history of blackouts around the world. There have been several blackouts around the world that affected more people than the northeast blackout did, although in many cases the amount of load lost was significantly less. The biggest blackout recorded so far in the world occurred on July 31, 2012, in India. Approximately 620 million people were affected in northern India. An outage in the Northeast (United States and Canada) in 1965 affected about thirty million people and led to the beginning of NERC (see chapter 4). The worst blackout in Italy's history occurred in September 2003, affecting over fifty million people. I remember this blackout since it occurred soon after the northeast blackout.

On July 13, 1977, there was a blackout in New York City that knocked out power for about eight million people. The chaos that ensued in New York for the next twenty-four hours is one of the reasons that operators have always looked at blackouts so seriously. They can be frightening events, especially when little or no information about what is going on is available to the public. People's lives are disrupted to a huge extent. It is not an exaggeration to say that a blackout is potentially life-threatening. This is why I have devoted a chapter to one such event, just to examine how it can happen.

There were also two partial grid blackouts in the United States in our western interconnection in the summer of 1996 that were important from the standpoint of a review of reliable operations. There was one on July 2 that affected over two million people, and another on August 10 that affected over seven million people in the United States and Canada. These outages led to the development of a new oversight role (that of reliability coordinator), as discussed in detail in chapter 2.

The Northeast Blackout of 2003

Those of us in the business of preventing blackouts as we operate this massive machine called the grid have spent a lot of time reviewing what happened to cause the northeast blackout. When millions of people are affected, it is a very big deal—and it is worthy of much study to understand why it happened. Initially there were people claiming it was an act of terrorism, or the result of the new competitive structure on the grid, or even the result of our grid's being too old. Well, we know that terrorism had nothing to do with this one. We also know that there were plenty of mistakes made to go around, and frankly, the newest grid in the world would have collapsed under the pressure. And as far as competitive structure goes, I will say emphatically that it had little to do with this blackout.

There was a lot of misinformation going around after this event. Someone had taken one of the nighttime satellite pictures of the eastern United States and Canada and simply erased all the lights from a huge area in the Northeast. My staff engineers and I were shocked to see this picture turn up in various reports about the blackout. The picture was a hoax. I must admit that the picture was pretty spectacular-looking. The truth is, there were pockets of electricity that somehow survived the blackout. A true satellite picture of the region after this blackout (there is one out there) shows small areas of lights through much of the Northeast, although there are big swaths of dark areas. I remember that we found the hoax picture in a report that our public affairs folks had planned to use and that they had asked us to review. Another interesting thing we heard was that the operators north of us were able to take action to stop the spread of the blackout. I wish this were true, but unfortunately it is not. Once the cascade began, it moved too fast for an operator to react.

Setting the Stage for a Blackout

I will set the stage for the events of August 14, 2003. As discussed earlier, reliability coordinators (RCs) had been established throughout the United States and Canada. There were about twenty RCs in operation at the time, including the Midwest Independent System Operator (MISO) for the First Energy area. MISO had only been acting as RC since February 1, 2003, however. There were standards for how operations, planning, maintenance, and other related activities were to be performed, but these standards had no bite since there were no penalties for noncompliance. All we had for enforcement was peer pressure. At this point in time, those of us involved in the electricity business were well aware of the limitations of this model.

It became evident during the post-event review done by NERC that the details of operating and communicating between the transmission owners and the RCs had not been fully worked out. The structure was designed in such a way that the RC would have an overview of several transmission owners' systems and would also have the tools that would allow the RC to back up the analyses performed by the transmission owners. And as discussed in chapter 2, the RC would have defined authority to handle problems quickly. The interface between two different RCs was not as clearly established, although obviously the intent of establishing these RCs was so that system operators would be able to curtail brewing problems quickly.

Also, and crucial to the blackout, a lot of money was being invested into unregulated generation. The clearing of trees near transmission lines had been neglected for several years at some utilities, as I suspect these activities were curtailed in order to divert the money to a better return on investment. Trees in the right-of-way of some utilities' transmission were growing tall; under one key transmission line in First Energy, the trees were found after the event to be forty-two feet tall!

August 14, 2003, was a hot but typical summer day. It was the kind of day that makes operations people earn their keep. No all-time system peaks were set this day in the area of the blackout. There were some generating units out, and during the day (prior to the blackout) some units tripped off-line, but this was not unusual and no one was raising concerns as the day got under way. It was a Friday afternoon, and it looked as if everything was fine.

It Takes a Combination of Big Mistakes to Cause a Blackout

I am going to step through a sequence of events that occurred during the afternoon. All times given are after noon on the day of the blackout and are based on the NERC study of the event (see "Reports" under the list of references at the end of this book for more detail). The first bad sign of things to come occurred at 12:15, when the MISO computer failed to run the contingency-analysis application the RC needs to determine if the system is secure. They got it running for a while, but then they turned off the automatic-run function. The MISO contingency-analysis application didn't produce any useful output again until 4:04.

At 1:31 a large unit (612 Mw) in First Energy's system tripped off-line. This in itself was not a huge problem for First Energy, but it did require more importing of energy. For a system that is bigger than 12,000 Mw, First Energy was importing 21 percent of its total load by 3:00. This put a strain on the transmission grid in the area, and voltages around Cleveland and Akron were sagging. But still there was no immediate emergency. Conditions necessitated close monitoring, though, and yet that didn't happen.

At 2:14, First Energy (FE) lost all its alarm and logging functions in the control room. The loss of alarms at FE was a key ingredient to this blackout, as not having this information, and not having any oversight from their RC, left FE in the dark as to what was actually happening. The problem was especially critical since these functions failed with no clear indication that they weren't working! When the alarm system fails, there are no alarms to warn the operators. The NERC report mentions that FE did not know for ninety minutes that the alarms were not working.

The unfortunate thing about the tools we use for situational awareness—the computers, data-gathering equipment, alarms, screens that show information to the operators, etc.—all seem to have a limit for how much they can deal with before they simply crash. My computer that I am typing on right now is no exception. These systems are tested to handle specified amounts of data, and they wouldn't be in service if they hadn't passed the test. But they seem to be most prone to failure when a lot is going on. And on the afternoon of August 14, there was a lot going on. As the afternoon progressed, transmission lines were tripping off-line and reclosing back, which resulted in a number of alarms. There were probably low-voltage alarms, alarms of flows approaching line limits (early warning), and alarms for equipment temperature approaching limits. All this was going on before the blackout. It probably overwhelmed the systems designed to keep the operators informed.

At 2:51 American Electric Power (AEP) operators called the FE operators to discuss a 345 kV tie-line trip and reclose that they saw on their monitoring equipment. Since the FE operators had no alarms, they believed the AEP information was incorrect. The FE operators had nothing in their control room to let them know that conditions on their system were slowly decaying. Many control rooms have a map board with some level of indication of the status of lines and switchyards in their part of the grid. In our control room, the map board acted as a backup to the computer system and allowed the operators to get a quick overview of the status of the system even if the computer and alarm functions were not operating. FE had no map board, no alarms, and no way to know that a line had tripped and reclosed, so they discounted what the AEP operators told them.

At 3:05, a 345 kV line tripped and locked out in FE's system. After this trip, NERC found that the FE system was not able to withstand several other contingent events. Of course, FE didn't know the line had tripped. Previously I have discussed the fact that when one transmission line trips off, the flows in all other lines increase to take up the slack. So after the trip of this line, all the lines in the area were getting higher loading. In many cases, the loading wasn't beyond limits at this point, although for certain contingencies the line limits would be exceeded. Now is when the problem of tall trees in the right-of-ways reared its head.

Within the next thirty-seven minutes, two additional 345 kV lines tripped and locked out. Interestingly, NERC found that the flows on these lines did not exceed their emergency rating. The problem is that the emergency rating was established based on an assumption that trees would be cleared under the conductors of the lines. Since trees were so high under the lines, there was not enough room for the sag to take place. So these lines tripped and locked out once they made contact with the trees.

Around this time, 3:30, the RCs in the area began discussing reducing flows on some of the major lines. Their approach was to use established procedures to curtail transactions that were shown to add flows to the lines of concern. Unfortunately, it was too little, too late. At this point in the event, the only way to salvage as much of the grid as possible was to do some severe load shedding. Based on taped conversations, the RCs were not using real-time information. In some cases they were discussing actions to take for potential contingent events for lines that were already tripped. It was also evident that the FE operators were beginning to realize that something was not right, although again it was too little, too late. At 3:42, the FE operator notified FE's computer-support people that the system did not seem to be working properly.

From 3:39 to 4:09, the 138 kV system began to cascade, as loss of a significant part of the 345 kV system put the load on the 138 kV lines. During this time, sixteen 138 kV lines tripped and locked out. Another major 345 kV line tripped at 4:05, and after this the cascade of the system was under way and could not be stopped. From this point, there is no reason for me to continue to investigate the details of how the blackout progressed. Recall, however, that there were parts of the grid that survived, including New England and northward, into parts of Canada—and, of course, everything in the South and West. In fact, to be precise, most of the eastern grid survived this event. I point this out to emphasize how resilient the grid can be.

Tools, Trees, Training: Lessons Learned

The biggest takeaways from this event were, sadly, similar to takeaways from previous, smaller blackouts. We called the takeaways the three t's: trees, tools, and training. The good news is that after this blackout, things were done that made a big difference in grid operators' joint ability to operate reliably. Standards were put into place that required better tree clearance, training, and tools for the operators. These standards were made mandatory, were backed up by huge fines, and most important, were enforced vigorously. The idea, which those of us in operations were only too aware of, was that any major player on the grid could make mistakes that lead to a blackout across a huge area. So the standards and enforcement were established to identify weak practices early on, prior to the next blackout. Nobody wanted to find instances of noncompliance only after the next blackout.

Hopefully, all transmission line rights-of-way today are maintained to have strict tolerances. Also, all the transmission operators, including the RCs, have been extensively trained in emergency operations. In fact, the standards require at least thirty-two hours of specific emergency training every year for all operators. My colleagues and I actually performed some of this training on a simulator, where we could present operators with situations that approached blackout conditions so they could practice preventing the collapse. The simulator was also useful for practicing recovery from a blackout. At Dominion we went a bit further and offered the blackout recovery simulation to nonoperators in the company to give them a better idea of what has to happen to put the system back together after a blackout. We thought that this may help our communications if we ever had to recover from a blackout.

All this improvement may make us think that such a blackout can't happen again. Unfortunately there have been a few smaller blackouts since the big one. In every case, lessons learned are shared throughout the industry. So why are there still blackouts occurring? Because we are only human and stuff breaks, I guess. Will a big one happen again? I would be a fool to say no. But hopefully the next one will be contained to a smaller area and recovery will be quicker.

CHAPTER 8

THE FUTURE OF THE ELECTRICITY BUSINESS

At midmorning on October 3, 1995, I was inside the control center when Dominion Virginia Power's system load started dropping off rapidly. The load dropped at least 300 Mw in a matter of just a few minutes. The frequency on the grid was running unusually high, especially for this time of day when load should be slowly increasing. What did I miss? Had we instituted a voltage reduction? Had we issued public appeals? Were we shedding load?

I asked one of the operators, "What in the world is going on?"

"Well, the verdict in the O. J. Simpson trial was just read by the jury," an operator said. "People are turning off their TVs."

Since I hadn't been following the trial, I wasn't prepared for the inevitable. It was bound to happen at some point, but how would anyone have been able to plan for this much load change due to thousands of TVs shutting off? Of course at the time all we could do was follow the load down by rapidly backing down generation. Will the future offer more efficient control alternatives to large load swings?

Introduction

In this chapter I explore several ongoing transformative drivers that have impacted, and will continue to impact, the grid. I have been asked many times if the grid will be necessary after an influx of distributed generation. In an effort to answer such a question, I will give my thoughts on the future of electricity transmission and distribution. Although it is highly doubtful, maybe someone will invent a way to beam energy from space, or invent a home-energy generator that fulfills everyone's energy needs without the grid. These kinds of technology breakthroughs would change everything, but I wouldn't count on them happening. There will be some significant changes, though, so I want to give my perspective.

The Death Spiral and Why It Won't Happen

There are those who forecast that the grid is doomed. The theory that I've seen in some articles seems to espouse the idea that with the onset of home generation, and other forms of distributed generation, the grid will become obsolete. The prediction is that as more of these innovations, such as roof-mounted solar generation units, are used, the grid will become a stranded investment. In this scenario, revenues to pay for the infrastructure of the grid will be falling precipitously as more and more customers go off the grid. Utilities will have to continuously raise rates in an attempt to recover their return on the investment, which will result in even more customers getting off the grid. Eventually, the owners of the grid will go bankrupt.

I want to closely examine this scenario with some technical and practical considerations. Assume that I install a solar panel on my roof with enough capability during summer months to net out all my energy needs during the day. In fact, the solar energy generated during the day exceeds the electricity demand in my household, so I start thinking about simply disconnecting from my local utility (i.e., going off the grid). I must decide what I am going to do about the nighttime load. I want to have some air-conditioning, have some lighting, and maybe watch television. Therefore, I am going to need some energy source when the sun isn't shining. I could buy a gasoline-powered home generator to fulfill that need, but that wouldn't be economic—nor would it be environmentally friendly.

Perhaps the best answer is to store the excess energy from the solar panels. The most practical energy-storage item today for a residence is a battery. How big does the battery need to be? That depends, of course. So I start thinking about the plan for getting the energy needed in the winter. I will have to have some way of generating energy in the winter, when the sun is not as giving as it is in the summer. The choice again is most likely a bigger battery for the summer that can last all winter. That isn't practical. These batteries are expensive, are big, and probably have a lifetime of five to ten years. They require maintenance and might be a fire hazard. I'm starting to get cold feet, but I will continue to explore alternatives.

In addition to the summer–winter energy-balance problem is the fact that in order to use the energy generated by the solar panels for many of my products, I am going to need an inverter to convert the direct current from the battery to the 60 Hz alternating current that most of my appliances need. I realize that the inverter is another device that can fail and leave me in the dark.

Another concern is having the capability to start motors such as an air-conditioner's compressor. These motors need a starting current that is as much as six times the normal running current. This means that the design of all of this equipment must take motor starting into account.

All this equipment will need to be oversized compared to the normal expected load, and this will add cost. In addition, all this stuff will need to be maintained so that it keeps working. By now I'm asking if the local utility will give me a deal on the energy I'm generating, because I see no benefit in separating from the grid.

Practically speaking, getting off the grid will be a new way of life for people, and frankly, the vast majority of people would not like it and will not do it. I'm sure there are people who are willing to forgo modern luxuries, but there are not enough of them to result in a grid death spiral. Some may try it for a while, but the maintenance, the equipment failures, the extra costs, and doing without things such as air-conditioning would eventually lead them back to the grid.

Most analyses I've seen on separating from the grid don't take into account the ability to start motors. And even these analyses conclude that separating is not cost-effective, which shows that the economics are potentially far worse than estimated. It will take some major advances in technology to change this outcome. Furthermore, this technology must be priced in a range that the masses can afford. I sincerely doubt that this is going to happen.

Clearly there are some in our midst who will opt to live off the grid. They may be willing to live with the hardships and/or to pay the extra cost. I just seriously doubt that it will be enough people to change the business model of owning a part of the grid.

Net Metering: Another Form of Subsidizing Renewables

There is an interesting nuance in the way many utilities are dealing with distributed generation today that needs some exploration. I refer to the net-metering approach that has been adopted in many states to incent the growth of renewables. The way this works is simple: I invest in a solar panel, I hook it up to my electric system behind the meter (connected on my side of the electric meter), and whatever my solar panels generate just nets out energy that I would otherwise be buying from the local utility. I'm still connected to the grid, and I get all the benefits of the energy available from the grid whenever the sun doesn't shine. I get the motor-starting capacity needed, I have energy when the solar panel/inverter system isn't working, and with the right kind of meter I can even sell excess energy to the utility.

In addition to these advantages, the grid provides all the energy-storage capability I need if I generate more than my own load, since the utility is paying me for the energy generated above my load. In fact, if I generate enough energy, my net bill could be zero, even though I have received all the benefits of being on the grid. I am in essence being subsidized by all the other rate payers who don't have solar panels. How does that work?

I mentioned above the benefit of the grid's acting as a huge energy-storage device for distributed generation (such as rooftop solar panels). Whatever is generated that is not needed locally can go back on the grid and be used elsewhere. I don't need to invest in and maintain a battery system, as the grid can take the energy. Earlier I said that the grid can't store energy, and that is a fact. But what it can do is absorb all the energy generated by these residential renewable systems by simply backing down other generators such as fossil fuel units.

As discussed in chapter 4, the typical residential customer pays an amount based on the total volume of energy used for the month. This is called *volumetric billing*, where customers pay for the kilowatt-hours used. Residential customers are not typically billed for a "demand charge," which is a fee for the infrastructure to deliver the total highest demand of the month. The idea is that the rate per kilowatt-hour is high enough to also cover the infrastructure charge for the distribution of the electricity. Utilities have estimated the total number of kilowatt-hours delivered for the year and have set the rate for those kilowatt-hours to cover the cost of the infrastructure with a return on investment. So with net metering and volumetric billing, all of these solar panels out there are avoiding the cost of the infrastructure. Those who don't have solar panels are making up the difference. How long will this go on?

The answer to that question lies in the political arena, but I dare say that the practice won't be allowed to continue until the grid owners go bankrupt. In fact, many states do not have a net-metering rate but instead would require a separate meter on the solar panels and pay the customer a fixed amount for the energy generated. Eventually, there may be a universal infrastructure charge for all residential customers that will cover the costs of the grid. This probably will not be based on a demand charge for each residential customer but will be a flat fee for connecting. Either way, those who enjoy the benefits of the grid will eventually be made to pay their share of the cost.

Microgrids and Other Ideas

Having shown the improbability of large numbers of residential customers separating from the grid, I would like to briefly address some other ideas about the future of the electric business. I'll start with the idea of microgrids. A microgrid is a combination of distribution equipment and multiple generators that could continue to operate if the local grid were not available (such as in a blackout). In the Lower 48, the idea may make some sense for critical facilities and possibly for industrial areas, although the cost to develop, maintain, and operate such a system would be high. (In remote areas such as Alaska, the microgrid may be a great solution, since there may be no grid nearby and few other options may be available.) Systems can be designed and programmed to automatically initiate a separation from the grid if it blacks out, and to start and run generation to match load.

Lots of issues would need scrutiny, however, especially in the case of joint ownership of the microgrid system. All of the equipment, especially the generators, will need to be maintained and tested occasionally or else the system will not be available when needed. The local grid owner will demand that the distribution system not feed power into his or her equipment, which is what we would call a *backfeed*. A backfeed can be a very dangerous situation for workers trying to recover the nearby portion of a failed grid. From a transmission operation perspective, we were always very hesitant to allow third-party entities to use our lines for their own deliveries after an outage, given the complication of recovering the system when pieces of the grid are being used by other parties.

The concept of a microgrid doesn't appear to be an approach to save energy dollars. Every way that I look at this, I see it is an extra cost. It would provide a potential degree of flexibility if the grid were lost, but how often does that actually happen? It may be a way to utilize renewable energy sources more directly than selling the output to the local utility, but I see this as more of a public relations move than as a move having economic value.

Similar to the microgrid concept is the idea of multiple homeowners and/or businesses effectively pooling their renewable energy outputs to better net out their energy demand. For example, customer A is not using the full energy output of his solar panels, but customer B is using more energy than her panels are producing, so the plan would be to have the extra energy from customer A routed to customer B. In effect, rather than selling the energy to the local utility used by customer A, it will just be directed to customer B's account. Since the local utility will pay less for energy than it will sell it for, the customers have found a great way to save some money.

First of all, this idea would require use of the local utility's lines (distribution). We call this *wheeling*. I say this because in reality the only way to get energy from customer A to customer B is to have the lines available that make the connection. There would have to be a wheeling fee charged in order to recover some of the cost of the infrastructure. Frankly, I doubt that any utility in the United States would go for this, since it would be extremely difficult to control all the flows. But one never knows what the future may hold.

The Smart Grid

No discussion of the future of the grid is complete without investigating the smart-grid concept. First of all, I have yet to see any universally accepted definition of smart grid. For purposes of simplicity, I'll just say that the idea is to better use information and automation to operate the electric system for the good of customers and of society in general. That said, there is a plethora of things that can be lumped under the heading of "smart grid." I'll get into a few just to introduce the subject and to help with my projections for the future of the grid. In this section I'm going to discuss the following things:

- the grid continues to get smarter, regardless of the term smart grid
- smart grid refers to distribution as well as transmission
- phasor measurement unit (PMU)
- better utilization of delivery lines
- advance metering infrastructure as a path to emission reductions
- load management
- self-healing distribution

The grid has been getting smarter ever since it was first developed. Lots of automation was used and improved on for a long time before anyone ever used the term *smart grid*. Those of us in the middle of the development were a bit baffled by the use of the term when it first came out. The term seemed to take on a new life when it was used by many who didn't really know what it meant. Eventually we had to realize that it was largely political (okay, so what isn't?). Sure, there are lots of things that can be done better, but there were times when it seemed that the idea was to make these improvements at any cost. Anyway, the push for the concept has helped get some government investment money (via the economic stimulus package) moving in the direction of the grid.

The term *smart grid* applies to all components of delivery of electricity from the power plant to the load. I have purposely focused discussion in this book on the transmission system (since it is mostly networked, as the term *grid* would imply), avoiding discussion of the lower-voltage distribution system because, for the most part, distribution is radial and not actually a grid as such. In my experience, I've found that almost anyone who talks about the benefits of the smart grid is referring to the distribution and transmission systems. So for this section, I will use the broadest definition that includes the distribution system.

I will begin the discussion by describing enhancements that improve operations at the system level. The first big-ticket item here is the phasor measurement unit (PMU) that will assist system operators to better determine impending trouble on the grid from a wide-area view. Hopefully, these devices will help to limit blackouts. This could be a big enhancement to situational awareness for the operators. PMUs provide important parameters of the delivery of large amounts of electricity over long distances so that the operators can know if the system is healthy.

In fact, information available before the 2003 northeast blackout using the PMU system would have shown that a blackout was imminent. An operator with this information (as well as the authority to act) could have prevented much of the cascade of the grid. These devices would also improve the performance of our analyses of system conditions in real time. The hard part is that in order for the PMUs to be of value, they must be deployed and coordinated over huge areas, such as the entire eastern interconnection.

Other enhancements that will come from the smart grid include more precise real-time capabilities of our transmission lines. Remember that the transmission lines are rated based on several factors, but one that is crucial is their thermal capability. Our ratings are generally based on an assumption of wind speed, since more wind blowing across the conductor will allow more current to flow, as the wind takes some of the heat away. Many of the calculated line ratings are based on an assumption of a wind speed of two miles per hour. But in most cases, a real-time measurement of wind speed would actually allow more current to flow. Since wind farms are producing their highest level of output during high winds, there could be times when the lines carrying their output could carry more than we would otherwise allow. So in this case, the information on wind speed in real time allows for increasing the rating on the lines, which allows for maximizing the output of the renewables.

Now I will get into the smart grid's impact on customers. Many utilities are installing metering equipment called *advanced metering infrastructure* (AMI), which is a big part of the smart-grid enhancement. These systems allow automated meter reading as well as remote cutoff or cut-on. The AMI will give outage information directly to the utility, so there will be no need for a customer to call when the lights are out. Getting this information directly will speed recovery as well.

These new meters will allow for remote reading of other information from the customer. For instance, the ability to read voltage at the customer end of the line allows for a tighter control of voltage in the distribution system. What has been found is that reducing voltage slightly on distribution lines actually results in less energy delivered. The savings in energy result in lowered carbon emissions, as well as monetary savings for the customer. However, there is a lower limit to the voltage. The AMI system allows for the tighter control without exceeding the limit.

The smart grid certainly includes many forms of load management, some of which have been seen for years. These programs, such as water-heating or air-conditioning control, will grow and can be even more effective with more real-time information. I have shown how load management can be used by system operators to reduce peak load in a system. But imagine how useful these controls could be to resolve a local overloading problem. In this case, instead of using the system-wide program, the load reduction can be specific to those customers whose loads impact the lines that have the problem.

Another term that is associated with the smart grid is *self-healing*. This refers to a grid that automatically detects and isolates electric faults and then reroutes power to the customers. Since the transmission part of the grid is networked, and since it has performed this function automatically for years, this discussion is for the distribution system. Just to be clear, many similar techniques have been used for years in the distribution system. Utilities can install a switch in the middle of a distribution line that only works if it detects a dead line. When that situation occurs, this switch works in tandem with other switches to isolate the part of the line that is damaged. The automatic switching then closes in to another source of power in order to restore service to many of the customers on the circuit. All this happens automatically and very quickly, so the customers may only have a momentary outage. The smart grid enhances these automatic operations with more real-time information, allowing for even more precise control of the power-delivery system.

As I said at the beginning of this section, there are a lot of smart-grid initiatives, some of which have been in place for years. I'm not going to attempt to address all of them. The result of these smart-grid enhancements will be an electric system that is more reliable and is operated to minimize emissions, at a cost to the customer that is less than it otherwise might have been. But please don't expect the smart grid to lower electric bills.

Cybersecurity

Way back in the early 1990s, I made a presentation in which I stated that the supervisory control and data acquisition (SCADA) system that utilities use for the actual control of grid devices was separate from the Internet and therefore impervious to cyberthreats. I wish now that I hadn't said it, but everybody makes mistakes! Today I'm afraid that the truth is far more complicated, with many unknowns. The Stuxnet attack on Iranian nuclear centrifuges in 2010 has changed my mind about the security of SCADA systems.

The smart-grid initiative, with all the exchanges of information and infrastructure that go along with it, has also opened up a risk to security. The people who run our information systems will admit that there are hundreds of illicit attempts for access every day. Every utility in the country is dealing with that as well. This raises concerns for the grid, especially if someone ever finds a way to hack into and get control of the grid equipment. There are lots of hackers who would do such a thing just to show they can. This gets scarier if it is someone with terrorist intent or if such a thing is done as an act of war.

Security of customer usage information is of paramount importance as these smart systems grow. Just think what a criminal could do with information that shows who isn't home at certain times of day. Also troubling is the fact that real-time cost data could show when the grid is most challenged, and this data will become available to all. Furthermore, if load and generation become more closely tied together based on the information system, the ties will offer more routes of disruption for the actors with ill intent.

It should be obvious by the opening statement in this section that I know very little about cybersecurity. There are NERC standards that address cybersecurity very specifically. These cyberstandards are complex. It takes specialists in the cybersecurity business to comprehend them. Additionally, cyberstandards have proven to be most often cited for noncompliance. This is clearly an area that justifies a lot of attention. I'm not suggesting that little is being done—just the opposite, in fact—but clearly the need for cybersecurity is here to stay.

Terrorism and the Grid

On April 16, 2013, several gunmen with high-powered rifles fired more than a hundred rounds into a substation on Pacific Gas and Electric's system in Silicon Valley. They were able to severely damage some transformers. As of this writing, they have not been caught. No one seems to know why they did this. Their act illustrates how vulnerable the electric grid is to attack. Our equipment is scattered over the countryside and, for the most part, is not guarded.

This is not the first case of vandalism on the grid, and it certainly won't be the last. But it got the attention of lots of people in Washington, DC (and elsewhere), who are very worried about the possibility of a terrorist attack on the US grid. Grid-level transformers are huge investments and take months or even years to replace, but they are unprotected all over the country. Some utility companies are building concrete walls around critical substations to protect against such an attack. But what can we do about the transmission towers and lines that cross thousands of miles?

The best thing we have going for us against any kind of physical attack is the resilience of the grid. There was a report a few years back that stated that shutting down fewer than ten critical substations in the eastern interconnection would cause the whole grid to collapse. Many grid operators doubted that the entire grid would completely black out from such an event, although certainly parts of it would. First of all, if a bad actor was intent on crashing the eastern interconnection, she or he would have to orchestrate a well-planned and coordinated attack at many sites. Even if the actor was successful in collapsing the entire interconnection, I have no doubt that portions of it could be recovered in a matter of hours. Nearly all of it could be recovered in a few days. Resilience of the grid was very evident in the northeast blackout of 2003. In that event, a huge portion of the eastern interconnection collapsed, but there were even larger portions of the eastern grid that remained energized.

At this point, I am more concerned about a concerted cyberattack on the grid than I am about a coordinated physical attack. The reason I feel this way is that the cyberattack can be conducted a long distance away, possibly from another country. There would be no need to coordinate teams of people trying to destroy equipment across large regions of the country. In a cyberattack, the nightmare is having our grid equipment suddenly doing strange things, like tripping off for no reason. A cyberattack would be difficult to respond to by the operations people, who wouldn't know initially why equipment was suddenly tripping off. But then again, there are multiple entities that control different parts of the eastern and western interconnection, and the bad actors would have to take over control at more than one large control center to be totally effective. It may be more likely in both the physical attack and the cyberattack scenarios that the intent is targeted to a specific area or a specific load as part of a larger scheme to cause disruption.

The effort to alleviate the impact of any terrorist attack is ongoing. All utility operators need to comply with the cyberstandards that have been developed to stave off the hackers and other bad actors. Also needed is a hardening of the most important substations throughout the country, without specifying to anyone which substations these are. The control centers need to be secured so that access is highly restricted and controlled. Ultimately, the training of operations staff to recognize and react appropriately to terrorist actions may be the best overall defense to limit the damage and the extent of an outage.

A Future View

What does the future electricity delivery system look like in the United States? At great risk of being wrong, I'm going to offer my view on the outlook for the grid up to about the year 2050, as follows:

- There will be three grids in the United States.
- There will be a reduction in the number of players.
- We will see increased alignment of goals for regulation.
- There will be improved coordination between load and generation on a huge scale.
- Future generation will be in nuclear, gas, and renewables.
- Customers will be able to choose an energy supplier.
- The grid will provide opportunities.

If my argument above, that the grid is not doomed to a death spiral, is convincing, then the first conclusion is that the grid will still be used and useful during the time frame under study. How many grids will there be in the United States? I predict that there will still be the grids we know today in the Lower 48. I don't see value in combining, or the political will to combine, any two grids into one. Texas will continue to operate as an independent (from FERC) entity, and it would simply cost too much to combine the east and west grids. Direct current ties, which may need to be supplemented, will continue to be an acceptable way to transfer power across the grid boundaries.

There are presently over three thousand entities that are in the electric business in some form in the United States. There are federally operated systems, municipal systems, investor-owned utilities, cooperatives, transmission-only companies, generation-only companies, and probably some I am leaving out. There are multiple entities that regulate this business, including the Federal Energy Regulatory Commission, most states, and municipalities. There are states that do not fall under FERC jurisdiction (namely, Texas, Alaska, and Hawaii). The electric business is fast approaching a time when wider-area coordination is the next step in achieving goals of improved reliability, reduced emissions, and lower cost for all customers. I predict that the number of players in this industry will be substantially reduced over time.

Uncertainty of regulation and differing state policies are both acting as a hindrance to progress in setting and achieving national goals. Frankly, one set of rules, and significantly fewer players, would make it a lot easier to move toward a more efficient future. So the question to ask is, how can we untangle ourselves from this mess to achieve better coordination in planning and operation? Our country has immense capability to achieve something when we all agree upon the goal. But as anyone who has watched our Congress perform over the last few years knows, there has been little incentive to work together. It is unclear how my vision for the energy industry can evolve unless there is a driving national concern. Climate change could be that driver over the next thirty-five years.

Improved coordination between load and generation, across state lines, can make some significant reductions in cost and emissions. Systems will be developed on a huge scale that will essentially use all renewable energy when it is available by coordinating generation directly with load. By doing this we will find that the use of peaking fossil fuel generators is not necessary as often, which will lower cost and emissions while at the same time making the fossil fuel units a bit more reliable thanks to less wear and tear.

Of course, this is not a new thing for the industry. Utilities have been doing load-management programs for years, longer than I've been in the business. But what will develop is a complex mechanism that will control an amount of load equivalent to the amount of renewable generation, with control systems that link the two. So we will be going from load-management programs that may control, say, 10 percent of the connected load today to programs that coordinate 30 percent of the load with available renewable generation.

The challenge is to find acceptable methods of reducing load when the wind stops blowing or the sun doesn't shine. I doubt that the populace will accept a system that doesn't give choice to the customer. For example, a mandated law for central control of loads in the home or factory is not going to come about. I'm hoping that the outcome of the new international goal to reduce emissions doesn't result in lowering our standard of living but that, instead, it allows for innovative ways to maintain what we are used to. Hopefully by using novel systems based on the cost of electricity in real time, and automation at the home or factory, we can successfully apply renewables at a much higher penetration level and thereby achieve targeted reductions in emissions.

The homeowner will have a device like a computer (perhaps the meter itself) that can be programmed to follow prices and other forecast data in real time, enabling the consumer to make decisions about when to run appliances. For example, I decide that at a cost of x dollars per kilowatt-hour or higher, I no longer want to charge my car or dry my clothes. But I also decide that when time is running out, I'm willing to pay extra to get the car charged when I need it. In this example, the direction to the home computer is more complicated than to a simple cutoff point, and the computer may have forecasts of wind and sun patterns (or may be linked to the central control computer of the local grid operator) that enhance the operation such that the car gets charged when needed, at the least cost and with the least emissions. The same kind of relationship can be imagined for the HVAC system, where the controller is programmed to allow a five-degree deviation from the thermostat setting to avoid a cost higher than x dollars. With this concept, similar controls can be imagined for the refrigerator, the dryer, and many other loads in the home.

But take this interface one step farther. What if the forecast load in the home, based on the home computer that is controlling it, is made available to a central control device that can assess the total upcoming loads, compare them to upcoming generation, and react to that forecast? This level of coordination can be a big improvement over the way systems are controlled today. Although the cost of doing this today would be enormous and prohibitive, I suspect that in thirty-five years the cost will be much lower and justified. It would also require a level of coordination across large geographic areas that we have simply not seen. Part of the justification is the significant reduction in carbon emissions that would result, especially if renewable generation penetration is along the lines of 50 percent or more. I'll wager that there will be a big cost adder to carbon emissions in our future and that this alone will drive some greatly enhanced control methods.

So the grid is still in use in, say, 2050, but what about the generation fleet? No question that renewables will become a bigger player, perhaps generating 25 percent of the total energy usage while providing up to 50 percent of the capacity. There will be national policies for renewable penetration, and although regulation at the state level will not be abandoned, at least states will be a bit more aligned. Distributed generation will continue to grow, with tax credits providing incentives, but the net-metering approach will not continue for the reasons discussed above.

Unless there are some major technical breakthroughs, coal generation is doomed. There will be virtually no new coal generation added (in the United States). By 2050 there will be no more than a handful of coal units left in operation. The major driver will be climate change, although clean air seems to be the driver today. Clean coal technology just isn't going to work well enough to keep coal alive, as sequestration of the carbon dioxide will not be economic.

There will still be nuclear generation as the public realizes that the risks are worth the low cost and the zero emissions. Sometime over the next thirty-five years, there will be a resurgence of nuclear power (meaning new nuclear power plants), driven in part by the economics of carbon taxes. There may even be a time when the government starts to take spent nuclear fuel, as it is supposed to do now.

Natural gas generation will still be around to fill in when necessary, but the high cost of carbon dioxide emissions will make it one of the highest-cost producers. This assumes that fracking isn't banned in the meantime, given its potential environmental impacts. And, of course, hydro will be a great resource to help meet peak loads.

Customers will have the means to choose when and if to buy electricity, based on cost, as I described above. The question then is, will customers have a choice of energy suppliers? As discussed previously, the retail choice initiatives so far have not proven to be such a good idea (such as in California). However, once the customer is armed with real-time information and the simple control capability to deal with that information automatically, the onset of retail choice is inevitable. There will continue to be competition at the wholesale level under the (federal) jurisdiction of FERC, and retail competition under the jurisdiction of state regulators. What drives this in part is a movement that aligns the states toward the goal of retail choice. As long as states have different objectives on this point, retail choice does not appear to be practical. Once retail choice is fully implemented, it will be the market that drives new and better ways of coordinating this massive system we have developed.

It is the interconnectivity of the grid that makes this coordination between generation and load possible. The whole idea of running the low-cost renewables in one region of the country and then transferring that energy to another region where fossil generation can be backed down is available to our society because of the grid. With the right tools, grid operators can keep the load-generation balance and allow for these transfers. Better utilization of this capability will lead to lower-cost energy and fewer emissions, without a significant impact on our way of life. To me it is clear that the grid is a crucial part of the energy picture and will be around for a long time to come.

GLOSSARY OF TERMS

alternating current: Alternating current is described in more detail in chapter 1. Most of our electric system is based on electric flow that actually alternates back and forth, sixty times per second.

area control error (ACE): Area control error is used by balancing authorities to control generators in such a way that the output of all generators matches the load in the assigned area. More precisely, ACE is the real-time difference between actual net interchange (the sum of all the ins and outs on the tie-lines) and scheduled net interchange (how much flow is expected), plus the factors of frequency bias and meter error.

automatic protection systems: These are extremely fast-acting systems that protect our grid from excessive damage if a fault occurs.

blackout: This term is used loosely but generally refers to a large area of load losing electric supply. It could also be applied to an entire grid collapsing, which is very rare. The blackout of 2003 only affected a portion of the eastern interconnection.

brownout: This is another term used loosely, and not always with consistency. For the most part when I see the term *brownout* used, it refers to when the load is purposefully removed because of problems on the grid, which is a localized event. I preferred the term *rotating blackout* for this event, never using *brownout* in this book.

capacity: The ability of a generator to produce power or of a transmission line to carry power. Capacity is measured in megawatts at the grid level.

cascading event: Once this event starts, it continues like a row of dominoes. A cascading event can start when one transmission line trips off-line and the load that was on that line is transferred to all the other lines in the area. Then another line becomes overloaded and trips.

combined-cycle gas generation: A combined-cycle unit will use waste-heat recovery after the initial gas combustion to increase efficiency.

contingency: An occurrence that results in the unexpected removal of some element from the grid, such as a transmission line or a generator.

cost-based: A regime where decisions are made based on the actual cost of the various options. This is opposed to a market-based or price-based regime, where decisions are made based on prices offered by marketers.

current or **amperes:** The actual flow of electricity is measured in amperes and can be called current. It usually flows in a conductor, but it could flow through a fault, such as a tree leaning on a transmission line. Physicists tell us that the individual electrons don't actually move very far, but the effect (the current) travels at the speed of light (I've never tested this).

demand: The load that is connected to the electric system (and drawing power/energy) at any given instant of time. Demand or load is usually measured in megawatts.

direct current: As the term implies, this is an electric source where the current only flows from one terminal to the other, as opposed to alternating current, where the current flows back and forth sixty times per second.

direct current tie-line: A transmission line that operates on direct current, ties two grids together, and does not require synchronism between the two grids.

dispatch: This refers to the controlling of generator output from a central controller that is remote from the generator.

distributed generation: This term is also used loosely, but it generally refers to generators that are on the distribution system, close to the load, typically small, and often not controlled by a central controller.

distribution: A delivery system that takes power off the transmission grid and distributes it to end-use customers. It typically includes voltages below 69 kV and is most often radial.

electric grid: This is the large interconnected system that takes power from generators and delivers it to the loads. In the United States, there are three somewhat separate grids. A grid by the definition that seems most often used is connected in such a way that the generators are all synchronous. The largest grid covers the entire East Coast, from Maine to Florida, and everything from there to the Rocky Mountains. The western interconnection is the second largest, covering Washington to California and everything eastward out to the Rockies. Then there is the Texas interconnection, which is located only in Texas and covers most of the state. Most of Canada and some parts of Mexico are also included in these grids.

energy: According to physics, energy is the ability to do work. In our terms, power (kilowatts) delivered for a length of time is energy (energy = power × time).

fault: A term referring to a short-circuit in the electric system, or an open circuit, that is caused by any number of things such as breaks, trees lying on lines, conductors lying on the ground, etc.

FERC: Federal Energy Regulatory Commission; see chapter 4 for an explanation of FERC's jurisdiction.

flowgate: Defined elements in the grid that are used by the North American Electric Reliability Corporation to calculate the flow through them caused by specific transactions. Typically a flowgate represents the weakest link in a critical area of the grid.

frequency: The number of times per second that a complete cycle occurs, such as sixty times per second in US grids. In such a case, the frequency is shown as 60 Hz.

frequency bias: A setting in the control algorithm for calculating area control error that allows for a boost to the grid from the balancing area for low or high frequency. See chapter 2 for more detail.

generator: A machine that converts other forms of energy into electrical energy. The generating unit consists of many components, one of which is the actual generator, which is turned by the turbine and produces electricity.

governor: The device/control within a generator that seeks to control the frequency output.

grid collapse: This is the worst of all outcomes, when the entire grid can't support delivery of electrical power anymore and all generation sources and loads are shut off. See also blackout.

grid reliability: The reliability of the transmission grid to serve its function of delivering the output of generators to distribution centers. This reliability is generally pretty high, as most outages for loads are in the distribution system.

hertz: The term given to the number of complete cycles of an alternating current in a second. For the US grid, sixty cycles per second is 60 Hz.

independent: For the purpose of the subject matter here, this term means that the entity that is controlling the market or transmission access is not affiliated with any owners of generation or transmission.

interchange: The instantaneous power flow on a tie-line between two entities on the grid, such as between two balancing authorities.

load: The connected devices that are consuming energy from the grid.

load following: The ability of a generator to increase or decrease its power output based on load. The control for load following would come from the central operations center that does generator dispatching.

load management: Any program that incents customers to allow the central controller to implement reductions in load when needed.

locational marginal pricing (LMP): A methodology to incent generation at locations on the grid that are not constrained by increasing the price paid for energy while reducing the price paid for energy on the wrong side of the constraint. The result is to reduce flows through an element on the grid that is near or over a limit.

locked out: This is the situation where circuit breakers on the grid open to prevent further damage during a fault, and will not reclose automatically. The circuit breakers cannot be closed again without a technician intervening.

loop flow: A natural occurrence in an interconnected grid where current will flow on multiple paths regardless of the contracted path. Loop flow refers to the flows that are not on the contract path.

losses: Energy that is dissipated in the delivery system due to resistance of the components that carry current. Loses are generally around 7 percent of the total delivered energy.

market-based pricing: Pricing that is based on bids by multiple parties in a competitive environment.

megawatts: A measure of power (not energy) in an electrical system. A watt is 1 ampere × 1 volt, so a megawatt is one million of those.

microgrid: This is another loosely used term that can cover many situations, but mainly it implies a small piece of the existing grid (like a city) that can be separated from the grid if necessary and then can continue operating.

NERC: North American Electric Reliability Corporation; see chapter 4 for an explanation of NERC's role.

network: An electrical system where there are multiple paths for the power to flow to get to the load. This is opposed to a radial system, where the power has only one path to the load.

OASIS: The Open Access Same-Time Information System used by transmission providers to sell access to marketers.

phases (electric): The alternating current in the grid is actually delivered over three lines, each carrying current and voltage that is 120 degrees out of sequence with the other. See chapter 1 for more detail and a diagram.

power: Normally thought of as the rate of energy delivery; measured in watts, kilowatts, and, on the transmission grid, in megawatts. Not to be confused with energy, which is measured in kilowatt-hours and megawatt-hours.

radial: An electrical system where the power has only one active path to the load.

reactive power: In order to deliver real power in our alternating current grid, there must be reactive power. Reactive power does not provide real power or energy, but it is a necessary ingredient for the grid to deliver real energy.

reconfigure: Refers to the practice of rearranging a networked system to remove potential flow problems or to help control voltage.

redispatch: When the least-cost generation plan causes flow problems on the grid, the mix of generation must be changed. This change is called redispatch.

retail: Sales of energy for final-use customers.

rotating blackouts: Purposely disconnecting portions of the load due to severe problems on the grid can be called a rotating blackout. The reason the blackout is rotated is to avoid prolonged outages on circuits. See chapter 3 for more detail.

short-circuit: This term is used to describe a fault in an electric circuit. It is called "short" because the power does not go to the load but has found an easier path where system operators don't want it to go.

simple-cycle gas turbine: These machines burn natural gas through a process much like that used by an internal combustion engine in a car. There is no recovery of waste heat in a simple-cycle turbine. They are typically able to reach full speed and output quickly.

smart grid: A series of automation and use of information to improve the efficiency and reliability of the grid. See chapter 8 for more detail.

solar magnetic disturbance: Solar events that can cause a disruption on the grid. See chapter 2 for more information.

stability: The ability of a generator or the grid to survive an electrical fault or oscillation and to stay connected with minimal impact.

synchronous: The term used to describe the situation that occurs when electrical generators are in step with the rest of the grid. See chapter 1 for more detail.

system operations: A term used loosely to describe the operations involving a part of the grid, as opposed to operations at a generating plant or at a distribution center.

system operator: A NERC-certified operator at a system operations center who is directly involved in grid-related control for a given area.

tie-line: A transmission line (usually) that connects two balancing areas, two transmission-owning companies, or two different interconnections (direct current in this case).

transaction: In the wholesale electric business, this is a contract between a seller and a buyer for resale that may include energy, capacity, and/or renewable energy credits.

transformer: A device that converts voltage from one level to another, allowing interconnections between systems of different voltages, such as 230 kV to 500 kV.

transmission: Generally thought of as taking the output of generators and delivering it to distribution centers for delivery to users. Arbitrarily, this output could be any voltage level above 69 kV.

tripped off: The term *tripped off* refers to an electrical device that has been disconnected from the system; the device is "locked out" when a device that is tripped off will not be automatically reconnected.

unit: A generation plant is usually made up of several units, each of which generates power. The units are for the most part independent, such that each unit can produce power independent of the others at that plant.

voltage: The property of electricity that refers to the "push" available, like the water pressure in a hose.

wholesale: The sale of products of generation on the grid, not for final use but for resale.

Y2K: The year 2000, when some people thought that the entire infrastructure of our society would collapse on account of software glitches.

SUGGESTED READING AND REFERENCES

Books

Casazza, Jack, and Frank Delea. *Understanding Electric Power Systems.* New Jersey: John Wiley & Sons, 2010.

O'Donnell, Arthur J. *Soul of the Grid.* Lincoln, NE: iUniverse, 2003.

Reports

Kelley, Michael B. "The Stuxnet Attack on Iran's Nuclear Plant 'Far More Dangerous' Than Previously Thought." Accessed September 3, 2015. http://www.businessinsider.com/stuxnet-was-far-more-dangerous-than-previous-thought-2013-11.

MIT Energy Initiative. *The Future of the Electric Grid; An Interdisciplinary MIT Study.* Accessed September 3, 2015. http://web.mit.edu/mitei/research/studies/the-electric-grid-2011.shtml.

NERC Steering Committee. *Technical Analysis of the August 14, 2003, Blackout.* Accessed September 3, 2015. http://www.nerc.com/docs/docs/blackout/NERC_Final_Blackout_Report_07_13_04.pdf.

North American Electric Reliability Council. *2010 Special Reliability Scenario Assessment.* Accessed September 3, 2015. http://www.nerc.com/pa/RAPA/ra/Reliability%20Assessments%20DL/EPA_Scenario_Final_v2.pdf.

United States Department of Energy (USDOE). *Final Report on the August 14, 2003 Blackout in the United States and Canada.* Accessed September 3, 2015. http://energy.gov/sites/prod/files/oeprod/DocumentsandMedia/BlackoutFinal-Web.pdf.

United States Department of Energy, Energy Information Administration. *Annual Energy Outlook, 2015.* Accessed September 3, 2015. http://www.eia.gov/forecasts/aeo/electricity_generation.cfm.

Websites

American Wind Energy Association: http://www.awea.org.

Electric Power Research Institute: http://www.epri.com/Pages/Default.aspx.

Environmental Protection Agency Clean Air Act: http://www2.epa.gov/laws-regulations/summary-clean-air-act.

Geothermal Energy Association: http://www.geo-energy.org/.

Institute of Electrical and Electronic Engineers: http://www.ieee.org/index.html.

International Geothermal Association: http://www.geothermal-energy.org/.

North American Electric Reliability Corporation (NERC): www.nerc.com.

PJM Independent System Operator: http://www.pjm.com/.

Solar Energy Industries Association: http://www.seia.org/.

Transmission Forum: http://www.natf.net/.

United States Department of Energy, Energy Information Administration (EIA): www.eia.gov.

United States Environmental Protection Agency (EPA): www.epa.gov.

United States Federal Energy Regulatory Commission (FERC): www.ferc.gov.

Other

Tacoma Narrows Bridge collapse: https://www.youtube.com/watch?v=j-zczJXSxnw.

INDEX

[diag denotes diagram]

A

ACE (area control error), 49–50, 81, 179
actual flow path, 56*diag*
advanced metering infrastructure (AMI), 170
affiliated businesses, separation of, 87, 89
alarms (to operators), 80–81
Alaska, 37, 167, 175
alternating current (AC), 11–13, 12*diag*, 14, 14*diag*, 22*diag*, 31, 179
Ameren, 19
American Electric Power (AEP), 95, 159
American Transmission Company (ATC) LLC, 20
amperes, 180
Annual Energy Outlook 2015, 128
area control error (ACE), 49–50, 81, 179
Associated Electric Cooperative, Inc., 19
Atlantic Municipal Utilities, 20
automatic protection systems, 40, 64–66, 179

B

backfeed, 167
bad things (that can happen to grid), 25–26

balance error, 46, 49
balancing areas, 45
balancing authority (BA), 45–46, 47, 48, 49, 50, 51, 56, 57, 60, 79, 81, 82, 116
barehanding, 66
base-loaded unit, 117
bilateral contracts, 142, 143, 144, 148, 149, 151
bilateral tariffs, 143
blackouts
    1965, 41, 44, 64, 102
    1996, 55, 60
    2003, 1, 9, 25, 67, 91, 92, 98, 155, 156
    combination of big mistakes as causing, 158–160
    competition as harbinger of, 59
    definition, 179
    factors that can lead to, 30–31, 36, 79
    history of, 155–156
    and law of unintended consequences, 27
    lessons learned from, 160–161
    rotating blackouts, 69, 75–78, 147, 185
    setting the stage for, 157–158
    training about, 58
    trees as causal factors in, 66, 67
    as worse-case event, 25–26, 30

black-start capability, 110, 112–113
Bonneville Power Administration, 19
brownout, 75, 179
bundled transmission, 151

C

California, design of market in, 38–39, 61, 62, 178
California Power Exchange, 38
capacitors, 31, 32, 71, 147
capacity, 36, 49, 67, 81, 88, 96, 97, 108, 109, 110, 111, 118, 126, 129, 142, 143–144, 146–147, 148, 149, 150, 177, 179. *See also* generation capacity; solar electric capacity; transmission capacity; wind capacity
capacity-backed transactions, 143–144
carbon capture and storage (CCS), 119
carbon tax, 118, 177
cascading events, 31, 35–36, 180
circuits, 9, 76. *See also* short-circuits
Clean Air Act, 118
CO2 emissions, 118, 119, 130
coal generation, 118–121, 177
combined-cycle gas generation, 180
Commonwealth Edison, 95
communications
    with field personnel, 82
    required of generation operators, 63
    rules for, 103
competition, in electricity business, 37, 38, 59, 60, 61, 85, 86, 89, 100, 104, 140–153, 178
compressed air storage, 132
concentrating solar power (CSP), 128
construction cost, 107, 108
contingency, 71, 76, 78, 79, 89, 158, 180. *See also* double contingency; single contingency
contract path, 56, 56*diag*
contracts, laws of, 56
control areas, 45

control centers/control rooms, 68, 72, 75, 78, 79, 80, 81, 82, 89, 120, 131, 158, 159, 173, 174
cooperatives, 19–20, 101, 175
cost-based, definition, 180
cost-based pricing, 148, 149, 150
cost-based rate, 84
cost-based system, 139, 149
cost(s)
    compared to price, 145–146
    construction cost, 107, 108
    going-forward cost, 107, 109
    incremental cost. *See* incremental cost
    levelized cost, 107, 108–109, 128, 129
CSP (concentrating solar power), 128
current, 180. *See also* alternating current (AC); direct current (DC)
cyberattack, 173–174
cybersecurity, 171–172

D

death spiral, 163–165
demand, 5, 8, 16, 19, 60, 69, 70, 72–73, 74, 75, 76, 79, 91, 96, 99, 100, 102, 115, 122, 138, 146, 163, 166, 167, 180
demand charge, 102, 166
Department of Energy (DOE), 98, 99, 128
deregulation, 18, 37
design rules, 40–44
direct current (DC), 12, 13, 14, 14*diag*, 15, 22, 164, 180
direct current tie-line, 174, 180
dispatch, 70, 120, 145–146, 180. *See also* redispatch
distributed battery storage, 132
distributed generation, 108, 163, 165, 177, 181
distribution
    compared to transmission, 11
    definition, 181

self-healing distribution, 168, 171
distribution line, 2, 22, 23, 170, 171
DOE (Department of Energy), 98, 99, 128
doing nothing, as option for problems, 78–79
Dominion Virginia Power, 19, 26, 27, 80, 95, 97, 110, 116, 144, 149, 150, 161
double contingency, 54
Duke Power, 19
dumb old utility guys (DOUGs), 59

E

eastern grid, 2, 13, 17, 21, 26, 29, 33, 35, 43, 44, 45, 48, 49, 51, 60, 160, 173
economy-energy transactions, 143
EIA (Energy Information Agency), 98, 124, 128
electric grid, definition, 181
Electric Power Research Institute (EPRI), 104–105
Electric Reliability Organization (ERO), 92, 103
electric systems
    interconnection of, 7–9
    small electric systems. *See* small electric system
electrical phases, 21–23, 22*diag*, 23*diag*, 184
electricity, flow of through grid to home, 23–25
electricity business
    competition in. *See* competition, in electricity business
    future of, 162–178
emergency energy transactions, 143
emergency operations, training in, 161
energy
    definition, 181
    power compared to, 4–5
Energy Information Agency (EIA), 98, 124, 128

energy management system (EMS), 79–80, 81
Energy Policy Act of 1992 (EPAct), 85–86, 88
Energy Policy Act of 2005 (EPAct05), 85, 92–93, 103
energy storage, 131–132, 135, 164, 165, 166
Enron, 59
EPA (US Environmental Protection Agency), 118, 132
EPRI (Electric Power Research Institute), 104–105
equilibrium, 17, 32, 50
equipment failure, 25, 30, 31, 33–35, 164
ERO (Electric Reliability Organization), 92, 103
exempt wholesale generator, 85

F

farm(s)
    as collection of wind turbines, 108. *See also* wind farms
    as large installation of fuel cells, 108
faults, 34, 36, 64, 65, 93, 112, 127, 171, 181
FE (First Energy), 157, 158, 159, 160
Federal Energy Regulatory Commission (FERC)
    ability of to fine, 67
    as approving of bilateral contracts and bilateral agreements, 142, 143, 144, 149
    as approving of standards, 103
    as approving of tariffs, 152
    definition, 181
    as enforcer of NERC standards, 98
    as granting NERC role of establishing rules, 103
    jurisdiction of, 19, 99, 174, 175

as moving electric industry toward
implementing wholesale
competition, 37
Order 693, 85, 93–94
as ordering implementation of
open access, 59
Orders 888 and 889, 55, 85,
86–91
overview, 84–94
as permitting returns on
transmission, 62
as responsible for rates, 101
on sales of firm transmission paths,
150–151
FERC Order 693, 85, 93–94
FERC Orders 888 and 889, 55, 85,
86–91
fines, 65, 66, 67, 79, 90, 92, 93,
103, 161
First Energy (FE), 157, 158, 159, 160
Florida Power and Light (FP&L), 19,
64, 93
flowgate, 57, 58, 81, 181
flywheel storage, 132
forecast, 37, 41, 51, 70, 73, 81, 109,
115, 176–177
Forum (North American Transmission
Forum), 104
frequency, 3, 11, 14, 15, 16, 17, 28, 30,
32, 33, 44, 49, 50, 81, 141, 182
frequency bias, 47, 48, 50, 51, 91, 182
fuel cells, 108, 130–131

G

generation
best mix of, 133–139, 139*diag*
coal generation, 118–121, 177
combined-cycle gas generation, 180
cost-based vs. price-based system:
the market, 139
distributed generation, 108, 163,
165, 177, 181
energy storage, 131–132
fuel cells, 130–131

general terms and concepts,
107–113
geothermal generation, 106, 129–
130, 133, 135
green energy, 132–133
home generation, 163
hydro generation, 123–124
impact of on frequency, 17
introduction, 106–107
and load, 25, 28, 30, 31, 32–33,
37, 44–52
natural gas generation, 121–
122, 178
nuclear generation, 106, 113–117,
125, 177
oil generation, 106, 110, 118, 120,
122–123
redispatching generation, 42,
69–70
rules for balancing of, 44–52,
46*diag*
solar generation, 110, 128–129,
163–164, 166
three-phase generation, 22
use of term, 24
wind generation, 110, 124–
128, 130
generation capacity, 61, 62, 85, 96, 124,
147, 150, 151
generation loss, 31, 35
generation plant vs. unit, 107
generation transactions, 86, 88
generator owners, 85
generators, 3, 4, 5, 6, 13, 16, 18, 23,
24, 32, 34, 42, 47, 63, 182
Geothermal Energy Association, 129
geothermal generation, 106, 129–130,
133, 135
going-forward cost, 107, 109
government, role of, 83–105. *See also
specific agencies*
governor, 3, 4, 44, 48, 182
green energy, 132–133
grid collapse, 30, 39, 44, 57, 58,
107, 182

grid control
    introduction, 29–30
    process of, 40–67
    rationale for, 30–40
grid reconfiguration, 69, 70–71, 184
grid reliability, 40–67, 182
grid(s)
    definition, 11, 181
    evolution of as not random, 17–18, 36
    flow of electricity through, 23–25
    getting off of, 164–165
    map of, 2*diag*
    as not storing power or energy, 17, 25
    number of, 2
    ownership of, 17, 18–20
    resilience of, 173
    use of term, 11
    in your home, 3*diag*

H

Hawaii, 37, 175
hertz, 11, 182
home generation, 163
human error, 25
hydro generation, 123–124
Hydro Quebec grid, 39

I

IEEE (Institute of Electrical and Electronics Engineers), 104
incremental cost, 38, 107, 109, 113, 115, 118, 127, 134, 135, 136, 137, 138, 140, 142, 143, 145
independent organizations, markets managed by, 152
independent system operators (ISOs), 94–97, 145–146, 147, 152
independents
    definition, 182
    generation or transmission built by, 152–153
inrush current, 64

Institute of Electrical and Electronics Engineers (IEEE), 104
interchange, 81, 182
interconnecting, 7–9, 13–15, 16–17, 16*diag*
interconnection agreements, 88
International Geothermal Association, 130
investor-owned utilities (IOUs), 18–19, 20, 45, 95, 98, 100, 152
ISOs (independent system operators), 94–97, 145–146, 147, 152

L

law of unintended consequences, 26–27
laws of contracts, 56
laws of physics, 18, 33, 45, 56, 107
levelized cost, 107, 108–109, 128, 129
lightning strikes, 65
LMP (locational marginal pricing), 96, 183
load
    definition, 182
    as devices and equipment, 3, 33
    electricity as going toward in radial system, 9
    generation and, 25, 28, 30, 31, 37, 44–52
    generators connected together with, 7*diag*
    grid as connecting generator to, 4
    percentage of as served by IOUs, 19
    rotating load curtailments, 39, 75
    rules for balancing of, 44–52, 46*diag*
    for typical shoulder season day, 137*diag*
    for typical summer peak day, 136*diag*
    for typical winter peak day, 137*diag*
    in water barrel analogy (for grid), 16*diag*
load balance, 31, 32–33, 48

load following, 110, 111, 113, 115, 116, 119, 121, 124, 125, 127, 130, 135, 138, 146, 147, 183
load management/load-management programs, 69, 72–73, 77, 141, 168, 170, 175, 176, 183
load shed/load shedding, 32, 33, 41, 42, 69, 75, 77, 79, 91, 97, 101, 151, 160
load-management agreements, 73
locational marginal pricing (LMP), 96, 183
location-based payments, 62
locked out, 159, 160, 183
loop flows, 57, 90, 91, 153, 183
loss of off-site power, 117
losses, 15, 54, 183

M

maintenance, 34, 40, 53, 66–67, 69–79, 90, 92, 97, 100, 103, 109, 157, 164
market power, 84, 149
market-based pricing, 148, 149, 183
markets
    competition, 140–153
    cost-based vs. price-based system, 139
    role of, 37–39
    rules for, 38, 59–63, 97
megawatts, definition, 183
microgrid, 166–167, 183
Midwest Independent System Operator (MISO), 157, 158
minimum output, 110, 111, 115, 116, 137
minimum take of energy, 144
municipalities, role of, 20

N

National Association of Regulatory Utility Commissioners (NARUC), 101

National Emission Standards for Hazardous Air Pollutants (NESHAP), 118
National Interest Electric Transmission Corridor, 99
natural gas generation, 121–122, 178
net metering, 165–166, 177
network system, 9–11, 57, 184
networked system, 9, 10*diag*, 56, 65
North American Electric Reliability Council (NERC)
    certification of as ERO, 92
    certification of operators by, 58
    definition, 184
    formation of, 41, 155
    overview, 102–104
    rules as developed by, 44, 54, 62, 63, 75, 79
    standards as established by, 50, 54, 84, 93–94, 98, 172
    study of 2003 blackout by, 157, 158, 159
North American Transmission Forum (Forum), 104
nuclear generation, 106, 113–117, 125, 177
Nuclear Regulatory Commission (NRC), 113–114

O

off-line, 42, 43, 52, 53, 114, 116, 117, 120, 123, 127, 157, 158, 159
off-line studies, 52
off-site power, loss of, 117
oil generation, 106, 110, 118, 120, 122–123
online, use of term, 53
Open Access Same-Time Information System (OASIS), 87, 89, 97, 150, 184
Open Access Transmission Tariff, 87, 146, 148–152
operating rules, 40, 44–58, 135
operator controls, 68–79

operators
    information available for, 79–82
    use of term, 79
overloads, 4, 9, 35, 36, 41, 52, 53, 57, 69, 70, 76, 78, 96, 171

P

Pacific Gas and Electric, 172
path of least resistance, 18, 24, 33, 39, 56
phases (electric), 21–23, 22*diag*, 23*diag*, 184
phasor measurement unit (PMU), 168, 169
photovoltaic (PV) generation, 128
physics, laws of, 18, 33, 45, 56, 107
PJM, 95–96, 97
planned outages, 34, 54, 58, 63, 66, 114
planning issues, 41–43
pooling, of renewable energy outputs, 167–168
power
    compared to energy, 4–5
    definition, 184
power flow, 5, 18, 45–46, 54, 80
power marketing administrations (PMAs), 19
price, cost compared to, 145–146
price-based system, 139
"Promoting Wholesale Competition through Open Access Non-discriminatory Transmission Service by Public Utilities," 86
protection systems, 64–66
public appeals, 69, 72, 74–75, 77, 91, 162
Public Utility Regulatory Policies Act of 1978 (PURPA), 86
pumped storage, 43, 132
PV (photovoltaic) generation, 128

R

radial system, 9, 10*diag*, 11, 64, 184
ramp rate, 110, 127

rate making, 99, 100, 101–102
RC (reliability coordinator), 58, 60, 82, 95, 156, 157, 160, 161
reactive capability, 110, 111, 126, 147, 149
reactive power, 31–32, 35, 37, 62, 63, 70, 71, 80, 111, 119, 146, 147, 184
reconfiguration (of grid), 69, 70–71, 184
"Recovery of Stranded Costs by Public Utilities and Transmitting Utilities," 86
redispatch
    definition, 184
    of generation, 53
    use of term, 70
redispatching generation, 42, 69–70
refueling/refueling outages, 114
regional entities, 55–58
regulation control, 47
reliability coordinator (RC), 58, 60, 82, 95, 156, 157, 160, 161
renewable energy, 129, 131, 133, 134, 165–166, 167, 175, 177
renewable energy certificates/credits (RECs), 125–126, 133, 146, 147
reserve issue, 62
resistance, path of least, 18, 24, 33, 39, 56
retail electricity, 18, 37, 184
right-of-way, clearing trees along, 66–67, 90–91, 160–161
rotating blackouts, 69, 75–78, 147, 185
rotating brownouts, 75
rotating load curtailments, 39, 75
rules
    in areas managed by ISOs, 152
    for balancing of generation and load, 44–52, 46*diag*
    for communications, 103
    design rules, 36, 40–44
    as important for grid reliability, 58–67
    for maintaining rights-of-way, 67

market rules, 38, 59–63, 97
NERC as responsible for, 102
operating rules, 40, 44–58, 91, 135
for reducing load, 75
reliability rules, 92, 102
states as tasked with establishing of regarding retail load, 100
for Texas grid, 15
under which utilities serve retail load, setting of, 100
Rural Electrification Act of 1936, 19

S

SCADA (supervisory control and data acquisition), 171
security, 171–172
SEIA (Solar Energy Industries Association), 129
self-healing distribution, 168, 171
series and parallel circuits, 9
short-circuits, 33–35, 43, 64, 185
simple-cycle gas turbines, 122, 136, 138, 185
simple-cycle units, 121
single contingency, 41, 42, 44, 52–54, 63, 70, 71, 81, 90, 97, 111, 147
situational awareness, 158–159, 169
small electric system
diagram of, 3*diag*, 6*diag*, 7*diag*
growth of, 5–7
overview, 3–4
smart grid, 11, 128, 132, 168–171, 172, 185
solar array, 108
solar electric capacity, 129
Solar Energy Industries Association (SEIA), 129
solar generation, 110, 128–129, 163–164, 166
solar installation, 108
solar magnetic disturbances (SMDs), 39–40, 43–44, 185
Southeastern Power Administration, 19
Southwestern Power Administration, 19

Special Reliability Scenario Assessment (2010), 118
stability, 31, 33, 36–37, 40, 41, 110, 112, 127, 131, 185
stacks, as large installations of fuel cells, 108
standards, 19, 41, 49, 52, 54, 58, 64, 84, 92, 93–94, 98, 103–104, 110, 116, 118, 134, 155, 157, 161, 172
start-up time, 110, 115, 116, 119–120, 122, 127
state regulation, 98–102
storms, 30, 31, 33–35, 73, 78
substations, 9, 10, 16, 21, 23, 30, 34, 40, 52, 76, 78, 80, 98, 104, 117, 126, 147, 173, 174
supervisory control and data acquisition (SCADA), 171
synchronicity, 13–15
synchronous, 13, 14, 15, 36, 112, 185
system design, 31, 35–36
system operations, 114, 115, 139, 185
system operator tools, 68–79
system operators
definition, 185
independent system operators (ISOs), 94–97, 145–146, 147, 152

T

Tacoma Narrows Bridge failure, 36
tariffs
bilateral tariffs, 143
Open Access Transmission Tariff, 87, 146, 148–152
Tennessee Valley Authority (TVA), 19
terrorism, 156, 172–174
Texas, 15, 37, 45, 84, 142, 174, 175
tie-lines, 7, 8*diag*, 14*diag*, 15, 19, 26, 142, 185. *See also* transmission lines
TOP (transmission operator). *See* transmission operator (TOP)

training
- about blackouts, 58
- in emergency operations, 161
- for implementation of plans, 112
- in operation of system, 128
- to recognize and react appropriately to terrorist actions, 174

transaction(s)
- capacity-backed transactions, 143–144
- definition, 186
- economy-energy transactions, 143
- emergency energy transactions, 143
- generation transactions, 86, 88
- wholesale electricity/wholesale transactions. *See* wholesale electricity/wholesale transactions

transformers, 1, 13, 21, 21*diag*, 23, 39, 43, 66, 125, 173, 186

transmission
- compared to distribution, 11
- definition, 186
- open access to, 87

transmission capacity, 86, 87, 89, 97, 150, 151

transmission investments, 62

transmission lines, 2, 10, 16, 18, 22, 23*diag*, 33, 41–42, 45, 52, 65, 66, 71, 98, 99, 169–170

transmission operator (TOP), 53, 54, 57, 58, 63, 70, 73, 79, 81, 82, 90, 91, 94, 104, 111, 114, 150, 151, 161

transmission-only companies, 20

trees, clearing of along right-of-way, 66–67, 90–91, 160–161

tripped off/tripping, 32, 33, 41, 116, 117, 186

tube leak, 118, 120–121, 123

TVA (Tennessee Valley Authority), 19

U

underfrequency load shedding, 32, 33

unintended consequences, 26–27, 87, 89–90

unit, definition, 186

United States–Canada Power System Outage Task Force, 98

unplanned events, 18, 34, 35, 52, 64, 65

US Congress, 55, 84–85, 95, 103, 175

US Constitution, on grid jurisdiction, 15

US Environmental Protection Agency (EPA), 118, 132

V

voltage, 20–21, 186

voltage collapse, 31, 32

voltage reduction, 69, 73–74, 75, 77

volumetric billing, 166

W

water barrel analogy (for grid), 16*diag*, 25

Western Area Power Administration, 19

western grid, 13, 57, 84

wheeling, 86, 89, 142, 143, 168

wholesale competition, 37, 59, 60, 85, 86, 104, 145

wholesale electricity/wholesale transactions, 37, 47, 55, 59–63, 102, 140–145, 146–148, 186

wind capacity, 124, 126

wind farms, 125, 126, 127, 128, 132, 170

wind generation, 110, 124–128, 130

Y

Y2K scare, 27, 106, 186

Made in United States
North Haven, CT
01 July 2024

54278138R00131